Numerical Solution of
Ordinary Differential Equations

Numerical Solution of Ordinary Differential Equations

Kendall E. Atkinson

Weimin Han

David Stewart

University of Iowa
Department of Mathematics
Iowa City, IA

WILEY

A JOHN WILEY & SONS, INC., PUBLICATION

Published by John Wiley & Sons, Inc., Hoboken, New Jersey.
Published simultaneously in Canada.

For general information on our other products and services or for technical support, please contact our Customer Care Department within the United States at (800) 762-2974, outside the United States at (317) 572-3993 or fax (317) 572-4002.

Wiley also publishes its books in a variety of electronic formats. Some content that appears in print may not be available in electronic format. For information about Wiley products, visit our web site at www.wiley.com.

Library of Congress Cataloging-in-Publication Data is available.

Atkinson, Kendall E.
  Numerical solution of ordinary differential equations / Kendall E. Atkinson, Weimin Han, David Stewart.
      p. cm.
  Includes bibliographical references and index.
  ISBN 978-0-470-04294-6 (cloth)
  1. Differential equations—Numerical solutions. I. Han, Weimin. II. Stewart, David, 1961- III. Title.
  QA372.A85 2009
  518'.63—dc22   2008036203

10 9 8 7 6 5 4 3 2 1

To Alice, Huidi, and Sue

Preface

This book is an expanded version of supplementary notes that we used for a course on ordinary differential equations for upper-division undergraduate students and beginning graduate students in mathematics, engineering, and sciences. The book introduces the numerical analysis of differential equations, describing the mathematical background for understanding numerical methods and giving information on what to expect when using them. As a reason for studying numerical methods as a part of a more general course on differential equations, many of the basic ideas of the numerical analysis of differential equations are tied closely to theoretical behavior associated with the problem being solved. For example, the criteria for the stability of a numerical method is closely connected to the stability of the differential equation problem being solved.

This book can be used for a one-semester course on the numerical solution of differential equations, or it can be used as a supplementary text for a course on the theory and application of differential equations. In the latter case, we present more about numerical methods than would ordinarily be covered in a class on ordinary differential equations. This allows the instructor some latitude in choosing what to include, and it allows the students to read further into topics that may interest them. For example, the book discusses methods for solving differential algebraic equations (Chapter 10) and Volterra integral equations (Chapter 12), topics not commonly included in an introductory text on the numerical solution of differential equations.

We also include MATLAB® programs to illustrate many of the ideas that are introduced in the text. Much is to be learned by experimenting with the numerical solution of differential equations. The programs in the book can be downloaded from the following website.

http://www.math.uiowa.edu/NumericalAnalysisODE/

This site also contains graphical user interfaces for use in experimenting with Euler's method and the backward Euler method. These are to be used from within the framework of MATLAB.

Numerical methods vary in their behavior, and the many different types of differential equation problems affect the performance of numerical methods in a variety of ways. An excellent book for "real world" examples of solving differential equations is that of Shampine, Gladwell, and Thompson [74].

The authors would like to thank Olaf Hansen, California State University at San Marcos, for his comments on reading an early version of the book. We also express our appreciation to John Wiley Publishers.

CONTENTS

Introduction

Differential equations are among the most important mathematical tools used in producing models in the physical sciences, biological sciences, and engineering. In this text, we consider numerical methods for solving ordinary differential equations, that is, those differential equations that have only one independent variable.

The differential equations we consider in most of the book are of the form

$$Y'(t) = f(t, Y(t)),$$

where $Y(t)$ is an unknown function that is being sought. The given function $f(t, y)$ of two variables defines the differential equation, and examples are given in Chapter 1. This equation is called a *first-order differential equation* because it contains a first-order derivative of the unknown function, but no higher-order derivative. The numerical methods for a first-order equation can be extended in a straightforward way to a system of first-order equations. Moreover, a higher-order differential equation can be reformulated as a system of first-order equations.

A brief discussion of the solvability theory of the initial value problem for ordinary differential equations is given in Chapter 1, where the concept of stability of differential equations is also introduced. The simplest numerical method, *Euler's method*, is studied in Chapter 2. It is not an efficient numerical method, but it is an intuitive way to introduce many important ideas. Higher-order equations and systems of first-order equations are considered in Chapter 3, and Euler's method is extended to such equations. In Chapter 4, we discuss some numerical methods with better numerical stability for practical computation. Chapters 5 and 6 cover more sophisticated and rapidly convergent methods, namely Runge–Kutta methods and the families of Adams–Bashforth and Adams–Moulton methods, respectively. In Chapter 7, we give a general treatment of the theory of multistep numerical methods. The numerical analysis of stiff differential equations is introduced in several early chapters, and it is explored at greater length in Chapters 8 and 9. In Chapter 10, we introduce the study and numerical solution of differential algebraic equations, applying some of the earlier material on stiff differential equations. In Chapter 11, we consider numerical methods for solving boundary value problems of second-order ordinary differential equations. The final chapter, Chapter 12, gives an introduction to the numerical solution of Volterra integral equations of the second kind, extending ideas introduced in earlier chapters for solving initial value problems. Appendices A and B contain brief introductions to Taylor polynomial approximations and polynomial interpolation.

CHAPTER 1

THEORY OF DIFFERENTIAL EQUATIONS: AN INTRODUCTION

For simple differential equations, it is possible to find closed form solutions. For example, given a function g, the general solution of the simplest equation

$$Y'(t) = g(t)$$

is

$$Y(t) = \int g(s)\, ds + c$$

with c an arbitrary integration constant. Here, $\int g(s)\, ds$ denotes any fixed antiderivative of g. The constant c, and thus a particular solution, can be obtained by specifying the value of $Y(t)$ at some given point:

$$Y(t_0) = Y_0.$$

Example 1.1 The general solution of the equation

$$Y'(t) = \sin(t)$$

is

$$Y(t) = -\cos(t) + c.$$

If we specify the condition

$$Y\left(\frac{\pi}{3}\right) = 2,$$

then it is easy to find $c = 2.5$. Thus the desired solution is

$$Y(t) = 2.5 - \cos(t). \qquad \blacksquare$$

The more general equation

$$Y'(t) = f(t, Y(t)) \tag{1.1}$$

is approached in a similar spirit, in the sense that usually there is a general solution dependent on a constant. To further illustrate this point, we consider some more examples that can be solved analytically. First, and foremost, is the first-order linear equation

$$Y'(t) = a(t)Y(t) + g(t). \tag{1.2}$$

The given functions $a(t)$ and $g(t)$ are assumed continuous. For this equation, we obtain

$$f(t, z) = a(t)z + g(t),$$

and the general solution of the equation can be found by the so-called *method of integrating factors*.

We illustrate the method of integrating factors through a particularly useful case,

$$Y'(t) = \lambda Y(t) + g(t) \tag{1.3}$$

with λ a given constant. Multiplying the linear equation (1.3) by the integrating factor $e^{-\lambda t}$, we can reformulate the equation as

$$\frac{d}{dt}\left(e^{-\lambda t}Y(t)\right) = e^{-\lambda t}g(t).$$

Integrating both sides from t_0 to t, we obtain

$$e^{-\lambda t}Y(t) = c + \int_{t_0}^{t} e^{-\lambda s}g(s)\, ds,$$

where

$$c = e^{-\lambda t_0}Y(t_0). \tag{1.4}$$

So the general solution of (1.3) is

$$Y(t) = e^{\lambda t}\left[c + \int_{t_0}^{t} e^{-\lambda s}g(s)\, ds\right] = ce^{\lambda t} + \int_{t_0}^{t} e^{\lambda(t-s)}g(s)\, ds. \tag{1.5}$$

This solution is valid on any interval on which $g(t)$ is continuous.

As we have seen from the discussions above, the general solution of the first-order equation (1.1) normally depends on an arbitrary integration constant. To single out a particular solution, we need to specify an additional condition. Usually such a condition is taken to be of the form

$$Y(t_0) = Y_0. \tag{1.6}$$

In many applications of the ordinary differential equation (1.1), the independent variable t plays the role of time, and t_0 can be interpreted as the initial time. So it is customary to call (1.6) an *initial value condition*. The differential equation (1.1) and the initial value condition (1.6) together form an *initial value problem*

$$\begin{aligned} Y'(t) &= f(t, Y(t)), \\ Y(t_0) &= Y_0. \end{aligned} \tag{1.7}$$

For the initial value problem of the linear equation (1.3), the solution is given by the formulas (1.5) and (1.4). We observe that the solution exists on any open interval where the data function $g(t)$ is continuous. This is a property for linear equations. For the initial value problem of the general linear equation (1.2), its solution exists on any open interval where the functions $a(t)$ and $g(t)$ are continuous. As we will see next through examples, when the ordinary differential equation (1.1) is nonlinear, even if the right-side function $f(t, z)$ has derivatives of any order, the solution of the corresponding initial value problem may exist on only a smaller interval.

Example 1.2 By a direct computation, it is easy to verify that the equation

$$Y'(t) = -[Y(t)]^2 + Y(t)$$

has a so-called trivial solution $Y(t) \equiv 0$ and a general solution

$$Y(t) = \frac{1}{1 + ce^{-t}} \tag{1.8}$$

with c arbitrary. Alternatively, this equation is a so-called separable equation, and its solution can be found by a standard method such as that described in Problem 4. To find the solution of the equation satisfying $Y(0) = 4$, we use the solution formula at $t = 0$:

$$4 = \frac{1}{1 + c};$$

$$c = -0.75.$$

So the solution of the initial value problem is

$$Y(t) = \frac{1}{1 - 0.75e^{-t}}, \qquad t \geq 0.$$

With a general initial value $Y(0) = Y_0 \neq 0$, the constant c in the solution formula (1.8) is given by $c = Y_0^{-1} - 1$. If $Y_0 > 0$, then $c > -1$, and the solution $Y(t)$ exists for $0 \leq t < \infty$. However, for $Y_0 < 0$, the solution exists only on the finite interval

$[0, \log(1 - Y_0^{-1}))$; the value $t = \log(1 - Y_0^{-1})$ is the zero of the denominator in the formula (1.8). Throughout this work, log denotes the natural logarithm.   ■

Example 1.3 Consider the equation

$$Y'(t) = -[Y(t)]^2.$$

It has a trivial solution $Y(t) \equiv 0$ and a general solution

$$Y(t) = \frac{1}{t+c} \qquad (1.9)$$

with c arbitrary. This can be verified by a direct calculation or by the method described in Problem 4. To find the solution of the equation satisfying the initial value condition $Y(0) = Y_0$, we distinguish several cases according to the value of Y_0. If $Y_0 = 0$, then the solution of the initial value problem is $Y(t) \equiv 0$ for any $t \geq 0$. If $Y_0 \neq 0$, then the solution of the initial value problem is

$$Y(t) = \frac{1}{t + Y_0^{-1}}.$$

For $Y_0 > 0$, the solution exists for any $t \geq 0$. For $Y_0 < 0$, the solution exists only on the interval $[0, -Y_0^{-1})$. As a side note, observe that for $0 < Y_0 < 1$ with $c = Y_0^{-1} - 1$, the solution (1.8) increases for $t \geq 0$, whereas for $Y_0 > 0$, the solution (1.9) with $c = Y_0^{-1}$ decreases for $t \geq 0$.   ■

Example 1.4 The solution of

$$Y'(t) = \lambda Y(t) + e^{-t}, \qquad Y(0) = 1$$

is obtained from (1.5) and (1.4) as

$$Y(t) = e^{\lambda t} + \int_0^t e^{\lambda(t-s)} e^{-s} \, ds.$$

If $\lambda \neq -1$, then

$$Y(t) = e^{\lambda t} \left\{ 1 + \frac{1}{\lambda + 1} [1 - e^{-(\lambda+1)t}] \right\}.$$

If $\lambda = -1$, then

$$Y(t) = e^{-t} (1 + t).$$ ■

We remark that for a general right-side function $f(t, z)$, it is usually not possible to solve the initial value problem (1.7) analytically. One such example is for the equation

$$Y' = e^{-t Y^4}.$$

In such a case, numerical methods are the only plausible way to compute solutions. Moreover, even when a differential equation can be solved analytically, the solution formula, such as (1.5), usually involves integrations of general functions. The integrals mostly have to be evaluated numerically. As an example, it is easy to verify that the solution of the problem

$$\begin{cases} Y' = 2tY + 1, & t > 0, \\ Y(0) = 1 \end{cases}$$

is

$$Y(t) = e^{t^2} \int_0^t e^{-s^2} ds + e^{t^2}.$$

For such a situation, it is usually more efficient to use numerical methods from the outset to solve the differential equation.

1.1 GENERAL SOLVABILITY THEORY

Before we consider numerical methods, it is useful to have some discussions on properties of the initial value problem (1.7). The following well-known result concerns the existence and uniqueness of a solution to this problem.

Theorem 1.5 *Let D be an open connected set in $\mathbb{R}^2$, let $f(t, y)$ be a continuous function of t and y for all (t, y) in D, and let (t_0, Y_0) be an interior point of D. Assume that $f(t, y)$ satisfies the* Lipschitz condition

$$|f(t, y_1) - f(t, y_2)| \leq K |y_1 - y_2| \qquad all \ (t, y_1), (t, y_2) \ in \ D \qquad (1.10)$$

for some $K \geq 0$. Then there is a unique function $Y(t)$ defined on an interval $[t_0 - \alpha, t_0 + \alpha]$ for some $\alpha > 0$, satisfying

$$Y'(t) = f(t, Y(t)), \qquad t_0 - \alpha \leq t \leq t_0 + \alpha,$$

$$Y(t_0) = Y_0.$$

The Lipschitz condition on f is assumed throughout the text. The condition (1.10) is easily obtained if $\partial f(t, y)/\partial y$ is a continuous function of (t, y) over $\overline{D}$, the closure of D, with D also assumed to be convex. (A set D is called *convex* if for any two points in D the line segment joining them is entirely contained in D. Examples of convex sets include circles, ellipses, triangles, parallelograms.) Then we can use

$$K = \max_{(t,y)\in\overline{D}} \left| \frac{\partial f(t, y)}{\partial y} \right|,$$

provided this is finite. If not, then simply use a smaller D, say, one that is bounded and contains (t_0, Y_0) in its interior. The number α in the statement of the theorem depends on the initial value problem (1.7). For some equations, such as the linear equation given in (1.3) with a continuous function $g(t)$, solutions exist for any t, and we can take α to be ∞. For many nonlinear equations, solutions can exist only in bounded intervals. We have seen such instances in Examples 1.2 and 1.3. Let us look at one more such example.

Example 1.6 Consider the initial value problem

$$Y'(t) = 2t[Y(t)]^2, \quad Y(0) = 1.$$

Here

$$f(t,y) = 2ty^2, \quad \frac{\partial f(t,y)}{\partial y} = 4ty,$$

and both of these functions are continuous for all (t, y). Thus, by Theorem 1.5 there is a unique solution to this initial value problem for t in a neighborhood of $t_0 = 0$. This solution is

$$Y(t) = \frac{1}{1 - t^2}, \quad -1 < t < 1.$$

This example illustrates that the continuity of $f(t, y)$ and $\partial f(t, y)/\partial y$ for all (t, y) does not imply the existence of a solution $Y(t)$ for all t. ∎

1.2 STABILITY OF THE INITIAL VALUE PROBLEM

When numerically solving the initial value problem (1.7), we will generally assume that the solution $Y(t)$ is being sought on a given finite interval $t_0 \leq t \leq b$. In that case, it is possible to obtain the following result on stability. Make a small change in the initial value for the initial value problem, changing Y_0 to $Y_0 + \epsilon$. Call the resulting solution $Y_\epsilon(t)$,

$$Y'_\epsilon(t) = f(t, Y_\epsilon(t)), \quad t_0 \leq t \leq b, \quad Y_\epsilon(t_0) = Y_0 + \epsilon. \tag{1.11}$$

Then, under hypotheses similar to those of Theorem 1.5, it can be shown that for all small values of ϵ, $Y(t)$ and $Y_\epsilon(t)$ exist on the interval $[t_0, b]$, and moreover,

$$\|Y_\epsilon - Y\|_\infty \equiv \max_{t_0 \leq t \leq b} |Y_\epsilon(t) - Y(t)| \leq c\epsilon \tag{1.12}$$

for some $c > 0$ that is independent of ϵ. Thus small changes in the initial value Y_0 will lead to small changes in the solution $Y(t)$ of the initial value problem. This is a desirable property for a variety of very practical reasons.

Example 1.7 The problem

$$Y'(t) = -Y(t) + 1, \quad 0 \leq t \leq b, \quad Y(0) = 1 \tag{1.13}$$

has the solution $Y(t) \equiv 1$. The perturbed problem

$$Y'_\epsilon(t) = -Y_\epsilon(t) + 1, \quad 0 \leq t \leq b, \quad Y_\epsilon(0) = 1 + \epsilon$$

has the solution $Y_\epsilon(t) = 1 + \epsilon e^{-t}$. Thus

$$Y(t) - Y_\epsilon(t) = -\epsilon e^{-t},$$

$$|Y(t) - Y_\epsilon(t)| \le |\epsilon|, \qquad 0 \le t \le b.$$

The problem (1.13) is said to be stable. ∎

Virtually all initial value problems (1.7) are stable in the sense specified in (1.12); but this is only a partial picture of the effect of small perturbations of the initial value Y_0. If the maximum error $\|Y_\epsilon - Y\|_\infty$ in (1.12) is not much larger than ϵ, then we say that the initial value problem (1.7) is *well-conditioned*. In contrast, when $\|Y_\epsilon - Y\|_\infty$ is much larger than ϵ [i.e., the minimal possible constant c in the estimate (1.12) is large], then the initial value problem (1.7) is considered to be *ill-conditioned*. Attempting to numerically solve such a problem will usually lead to large errors in the computed solution. In practice, there is a continuum of problems ranging from well-conditioned to ill-conditioned, and the extent of the ill-conditioning affects the possible accuracy with which the solution Y can be found numerically, regardless of the numerical method being used.

Example 1.8 The problem

$$Y'(t) = \lambda\,[Y(t) - 1], \qquad 0 \le t \le b, \qquad Y(0) = 1 \tag{1.14}$$

has the solution

$$Y(t) = 1, \qquad 0 \le t \le b.$$

The perturbed problem

$$Y_\epsilon'(t) = \lambda[Y_\epsilon(t) - 1], \qquad 0 \le t \le b, \qquad Y_\epsilon(0) = 1 + \epsilon$$

has the solution

$$Y_\epsilon(t) = 1 + \epsilon e^{\lambda t}, \qquad 0 \le t \le b.$$

For the error, we obtain

$$Y(t) - Y_\epsilon(t) = -\epsilon e^{\lambda t}, \tag{1.15}$$

$$\max_{0 \le t \le b} |Y(t) - Y_\epsilon(t)| = \begin{cases} |\epsilon|, & \lambda \le 0, \\ |\epsilon|\,e^{\lambda b}, & \lambda \ge 0. \end{cases}$$

If $\lambda < 0$, the error $|Y(t) - Y_\epsilon(t)|$ decreases as t increases. We see that (1.14) is well-conditioned when $\lambda \le 0$. In contrast, for $\lambda > 0$, the error $|Y(t) - Y_\epsilon(t)|$ increases as t increases. And for λb moderately large, say $\lambda b \ge 10$, the change in $Y(t)$ is quite significant at $t = b$. The problem (1.14) is increasingly ill-conditioned as λ increases. ∎

For the more general initial value problem (1.7) and the perturbed problem (1.11), one can show that

$$Y(t) - Y_\epsilon(t) \approx -\epsilon \exp\left(\int_{t_0}^{t} g(s)\,ds\right) \tag{1.16}$$

with

$$g(t) = \left.\frac{\partial f(t, y)}{\partial y}\right|_{y=Y(t)}$$

for t sufficiently close to t_0. Note that this formula correctly predicts (1.15), since in that case

$$f(t, y) = \lambda(y - 1),$$

$$\frac{\partial f(t, y)}{\partial y} = \lambda,$$

$$\int_0^t g(s)\, ds = \lambda t.$$

Then (1.16) yields

$$Y(t) - Y_\epsilon(t) \approx -\epsilon e^{\lambda t},$$

which agrees with the earlier formula (1.15).

Example 1.9 The problem

$$Y'(t) = -[Y(t)]^2, \qquad Y(0) = 1 \tag{1.17}$$

has the solution

$$Y(t) = \frac{1}{t + 1}.$$

For the perturbed problem,

$$Y'_\epsilon(t) = -[Y_\epsilon(t)]^2, \qquad Y_\epsilon(0) = 1 + \epsilon, \tag{1.18}$$

we use (1.16) to estimate $Y(t) - Y_\epsilon(t)$. First,

$$f(t, y) = -y^2,$$

$$\frac{\partial f(t, y)}{\partial y} = -2y,$$

$$g(t) = -2Y(t) = -\frac{2}{t + 1},$$

$$\int_0^t g(s)\, ds = -2 \int_0^t \frac{ds}{s + 1} = -2\log(1 + t) = \log(1 + t)^{-2},$$

$$\exp\left[\int_0^t g(s)\, ds\right] = e^{\log(t+1)^{-2}} = \frac{1}{(t + 1)^2}.$$

For $t \geq 0$ sufficiently small, substituting into (1.16) gives

$$Y(t) - Y_\epsilon(t) \approx \frac{-\epsilon}{(1 + t)^2}. \tag{1.19}$$

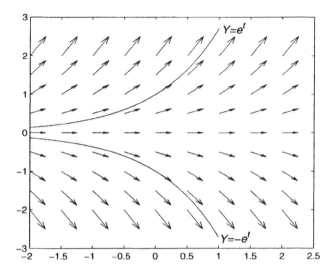

Figure 1.1 The direction field of the equation $Y' = Y$ and solutions $Y = \pm e^t$

This indicates that (1.17) is a well-conditioned problem. ■

In general, if

$$\frac{\partial f(t, Y(t))}{\partial y} \leq 0, \qquad t_0 \leq t \leq b, \tag{1.20}$$

then the initial value problem is generally considered to be well-conditioned. Although this test depends on $Y(t)$ over the interval $[t_0, b]$, one can often show (1.20) without knowing $Y(t)$ explicitly; see Problems 5, 6.

1.3 DIRECTION FIELDS

Direction fields serve as a useful tool in understanding the behavior of solutions of a differential equation. We notice that the graph of a solution of the equation $Y' = f(t, Y)$ is such that at any point (t, y) on the solution curve, the slope is $f(t, y)$. The slopes can be represented graphically in direction field diagrams. In MATLAB®, direction fields can be generated by using the meshgrid and quiver commands.

Example 1.10 Consider the equation $Y' = Y$. The slope of a solution curve at a point (t, y) on the curve is y, which is independent of t. We generate a direction field diagram with the following MATLAB code:
First draw the direction field:

```
[t,y] = meshgrid(-2:0.5:2,-2:0.5:2);
```

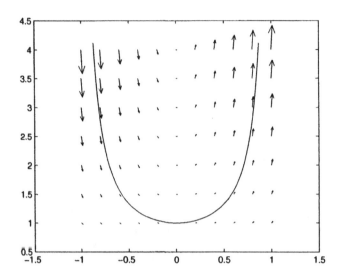

Figure 1.2 The direction field of the equation $Y' = 2tY^2$ and the solution $Y = 1/(1 - t^2)$

```
dt = ones(9);          %Generates a matrix of 1's.
dy = y;
quiver(t,y,dt,dy);
```
Then draw two solution curves:
```
hold on
t = -2:0.01:1;
y1 = exp(t); y2 = -exp(t);
plot(t,y1,t,y2)
text(1.1,2.8,'\itY=e^t','FontSize',14)
text(1.1,-2.8,'\itY=-e^t','FontSize',14)
hold off
```
The result is shown in Figure 1.1. ■

Example 1.11 Continuing Example 1.6, we use the following MATLAB M-file to generate a direction field diagram and the particular solution $Y = 1/(1 - t^2)$ in Figure 1.2.
```
[t,y] = meshgrid(-1:0.2:1,1:0.5:4);
dt = ones(7,11); dy = 2*t.*y.^2;
quiver(t,y,dt,dy);
hold on
tt = -0.87:0.01:0.87;
```

```
yy = 1./(1-tt.^2);
plot(tt,yy)
hold off
```

Note that for large y values, the arrows in the direction field diagram (Figure 1.2) point almost vertically. This suggests that a solution to the equation may exist only in a bounded interval of the t axis, which, indeed, is the case. ∎

PROBLEMS

1. In each of the following cases, show that the given function $Y(t)$ satisfies the associated differential equation. Then determine the value of c required by the initial condition. Finally, with reference to the general format in (1.7), identify $f(t, z)$ for each differential equation.

   (a) $Y'(t) = -Y(t) + \sin(t) + \cos(t)$, $Y(0) = 1$;
   $Y(t) = \sin(t) + ce^{-t}$.

   (b) $Y'(t) = \left[Y(t) - Y(t)^2\right]/t$, $Y(1) = 2$; $Y(t) = t/(t + c)$, $t > 0$.

   (c) $Y'(t) = \cos^2(Y(t))$, $Y(0) = \pi/4$; $Y(t) = \tan^{-1}(t + c)$.

   (d) $Y'(t) = Y(t)[Y(t) - 1]$, $Y(0) = 1/2$; $Y(t) = 1/(1 + ce^t)$.

2. Use MATLAB to draw direction fields for the differential equations listed in Problem 1.

3. Solve the following problem by using (1.5) and (1.4):

   (a) $Y'(t) = \lambda Y(t) + 1$, $\quad Y(0) = 1$.

   (b) $Y'(t) = \lambda Y(t) + t$, $\quad Y(0) = 3$.

4. Consider the differential equation

$$Y'(t) = f_1(t)f_2(Y(t))$$

for some given functions $f_1(t)$ and $f_2(z)$. This is called a *separable* differential equation, and it can be solved by direct integration. Write the equation as

$$\frac{Y'(t)}{f_2(Y(t))} = f_1(t),$$

and find the antiderivative of each side:

$$\int \frac{Y'(t)\, dt}{f_2(Y(t))} = \int f_1(t)\, dt.$$

On the left side, change the integration variable by letting $z = Y(t)$. Then the equation becomes

$$\int \frac{dz}{f_2(z)} = \int f_1(t)\, dt.$$

After integrating, replace z by $Y(t)$; then solve for $Y(t)$, if possible. If these integrals can be evaluated, then the differential equation can be solved. Do so for the following problems, finding the general solution and the solution satisfying the given initial condition.

(a) $Y'(t) = t/Y(t)$, $Y(0) = 2$.

(b) $Y'(t) = te^{-Y(t)}$, $Y(1) = 0$.

(c) $Y'(t) = Y(t)[a - Y(t)]$, $Y(0) = a/2$, $a > 0$.

5. Check the conditioning of the initial value problems in Problem 1. Use the test (1.20).

6. Check the conditioning of the initial value problems in Problem 4 (a), (b). Use the test (1.20).

7. Use (1.20) to discuss the conditioning of the problem

$$Y'(t) = Y(t)^2 - 5\sin(t) - 25\cos^2(t), \quad Y(0) = 6.$$

You do not need to know the true solution.

8. Consider the solutions $Y(t)$ of

$$Y'(t) + aY(t) = de^{-bt}$$

with a, b, d constants and $a, b > 0$. Calculate

$$\lim_{t \to \infty} Y(t).$$

Hint: Consider the cases $a \neq b$ and $a = b$ separately.

CHAPTER 2

EULER'S METHOD

Although it is possible to derive solution formulas for some ordinary differential equations, as is shown in Chapter 1, many differential equations arising in applications are so complicated that it is impractical to have solution formulas. Even when a solution formula is available, it may involve integrals that can be calculated only by using a numerical quadrature formula. In either situation, numerical methods provide a powerful alternative tool for solving the differential equation.

The simplest numerical method for solving the initial value problem is called *Euler's method*. We first define it and give some numerical illustrations, and then we analyze it mathematically. Euler's method is not an efficient numerical method, but many of the ideas involved in the numerical solution of differential equations are introduced most simply with it.

Before beginning, we establish some notation that will be used in the rest of this book. As before, $Y(t)$ denotes the true solution of the initial value problem with the initial value Y_0:

$$Y'(t) = f(t, Y(t)), \quad t_0 \le t \le b,$$
$$Y(t_0) = Y_0. \tag{2.1}$$

Numerical methods for solving (2.1) will find an approximate solution $y(t)$ at a discrete set of nodes,

$$t_0 < t_1 < t_2 < \cdots < t_N \leq b. \tag{2.2}$$

For simplicity, we will take these nodes to be evenly spaced:

$$t_n = t_0 + nh, \quad n = 0, 1, \ldots, N.$$

The approximate solution will be denoted using $y(t)$, with some variations. The following notations are all used for the approximate solution at the node points:

$$y(t_n) = y_h(t_n) = y_n, \quad n = 0, 1, \ldots, N.$$

To obtain an approximate solution $y(t)$ at points in $[t_0, b]$ other than those in (2.2), some form of interpolation must be used. We will not consider that problem here, although there are standard techniques from the theory of interpolation that can be easily applied. For an introduction to interpolation theory, see, e.g., [11, Chap. 3], [12, Chap. 4], [57, Chap. 8], [68, Chap. 8].

2.1   DEFINITION OF EULER'S METHOD

To derive Euler's method, consider the standard derivative approximation from beginning calculus,

$$Y'(t) \approx \frac{1}{h}[Y(t + h) - Y(t)]. \tag{2.3}$$

This is called a *forward difference approximation* to the derivative. Applying this to the initial value problem (2.1) at $t = t_n$,

$$Y'(t_n) = f(t_n, Y(t_n)),$$

we obtain

$$\frac{1}{h}[Y(t_{n+1}) - Y(t_n)] \approx f(t_n, Y(t_n)),$$

$$Y(t_{n+1}) \approx Y(t_n) + hf(t_n, Y(t_n)). \tag{2.4}$$

Euler's method is defined by taking this to be exact:

$$y_{n+1} = y_n + hf(t_n, y_n), \quad 0 \leq n \leq N - 1. \tag{2.5}$$

For the initial guess, use $y_0 = Y_0$ or some close approximation of Y_0. Sometimes Y_0 is obtained empirically and thus may be known only approximately. Formula (2.5) gives a rule for computing $y_1, y_2, \ldots, y_N$ in succession. This is typical of most numerical methods for solving ordinary differential equations.

Some geometric insight into Euler's method is given in Figure 2.1. The line $z = p(t)$ that is tangent to the graph of $z = Y(t)$ at t_n has slope

$$Y'(t_n) = f(t_n, Y(t_n)).$$

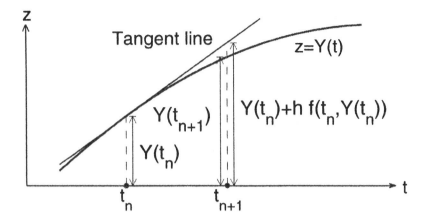

Figure 2.1   An illustration of Euler's method derivation

Using this tangent line to approximate the curve near the point $(t_n, Y(t_n))$, the value of the tangent line

$$p(t) = Y(t_n) + f(t_n, Y(t_n))(t - t_n)$$

at $t = t_{n+1}$ is given by the right side of (2.4).

Example 2.1   The true solution of the problem

$$Y'(t) = -Y(t), \qquad Y(0) = 1 \tag{2.6}$$

is $Y(t) = e^{-t}$. Euler's method is given by

$$y_{n+1} = y_n - hy_n, \qquad n \geq 0 \tag{2.7}$$

with $y_0 = 1$ and $t_n = nh$. The solution $y(t)$ for three values of h and selected values of t is given in Table 2.1. To illustrate the procedure, we compute y_1 and y_2 when $h = 0.1$. From (2.7), we obtain

$$y_1 = y_0 - hy_0 = 1 - (0.1)(1) = 0.9, \qquad t_1 = 0.1,$$
$$y_2 = y_1 - hy_1 = 0.9 - (0.1)(0.9) = 0.81, \qquad t_2 = 0.2.$$

For the error in these values, we have

$$Y(t_1) - y_1 = e^{-0.1} - y_1 \doteq 0.004837,$$
$$Y(t_2) - y_2 = e^{-0.2} - y_2 \doteq 0.008731.$$

∎

Table 2.1 Euler's method for (2.6)

h	t	$y_h(t)$	Error	Relative Error
0.2	1.0	3.2768e − 1	4.02e − 2	0.109
	2.0	1.0738e − 1	2.80e − 2	0.207
	3.0	3.5184e − 2	1.46e − 2	0.293
	4.0	1.1529e − 2	6.79e − 3	0.371
	5.0	3.7779e − 3	2.96e − 3	0.439
0.1	1.0	3.4867e − 1	1.92e − 2	0.0522
	2.0	1.2158e − 1	1.38e − 2	0.102
	3.0	4.2391e − 2	7.40e − 3	0.149
	4.0	1.4781e − 2	3.53e − 3	0.193
	5.0	5.1538e − 3	1.58e − 3	0.234
0.05	1.0	3.5849e − 1	9.39e − 3	0.0255
	2.0	1.2851e − 1	6.82e − 3	0.0504
	3.0	4.6070e − 2	3.72e − 3	0.0747
	4.0	1.6515e − 2	1.80e − 3	0.0983
	5.0	5.9205e − 3	8.17e − 4	0.121

Example 2.2 Solve

$$Y'(t) = \frac{Y(t) + t^2 - 2}{t + 1}, \qquad Y(0) = 2 \tag{2.8}$$

whose true solution is

$$Y(t) = t^2 + 2t + 2 - 2(t + 1)\log(t + 1).$$

Euler's method for this differential equation is

$$y_{n+1} = y_n + \frac{h(y_n + t_n^2 - 2)}{t_n + 1}, \qquad n \geq 0$$

with $y_0 = 2$ and $t_n = nh$. The solution $y(t)$ is given in Table 2.2 for three values of h and selected values of t. A graph of the solution $y_h(t)$ for $h = 0.2$ is given in Figure 2.2. The node values $y_h(t_n)$ have been connected by straight line segments in the graph. Note that the horizontal and vertical scales are different. ∎

In both examples, observe the behavior of the error as h decreases. For each fixed value of t, note that the errors decrease by a factor of about 2 when h is halved. As

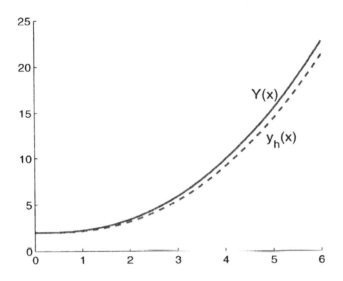

Figure 2.2 Euler's method for problem (2.8), $h = 0.2$

an illustration, take Example 2.1 with $t = 5.0$. The errors for $h = 0.2, 0.1$, and 0.05, respectively, are

$$2.96 \times 10^{-3}, \qquad 1.58 \times 10^{-3}, \qquad 8.17 \times 10^{-4}$$

and these decrease by successive factors of 1.93 and 1.87. The reader should do the same calculation for other values of t, in both Examples 2.1 and 2.2. Also, note that the behavior of the error as t increases may be quite different from the behavior of the relative error. In Example 2.2, the relative errors increase initially, and then they decrease with increasing t.

MATLAB® program. The following MATLAB program implements Euler's method. The Euler method is also called the *forward Euler method*. The *backward Euler method* is discussed in Chapter 4.

```
function [t,y] = euler_for(t0,y0,t_end,h,fcn)
%
% function [t,y]=euler_for(t0,y0,t_end,h,fcn)
%
% Solve the initial value problem
%     y' = f(t,y),   t0 <= t <= b,   y(t0)=y0
% Use Euler's method with a stepsize of h.  The user must
% supply a program to define the right side function of the
% differential equation.   Use some name, say deriv, and a
```

Table 2.2   Euler's method for (2.8)

h	t	$y_h(t)$	Error	Relative Error
0.2	1.0	2.1592	6.82e − 2	0.0306
	2.0	3.1697	2.39e − 1	0.0701
	3.0	5.4332	4.76e − 1	0.0805
	4.0	9.1411	7.65e − 1	0.129
	5.0	14.406	1.09	0.0703
	6.0	21.303	1.45	0.0637
0.1	1.0	2.1912	3.63e − 2	0.0163
	2.0	3.2841	1.24e − 1	0.0364
	3.0	5.6636	2.46e − 1	0.0416
	4.0	9.5125	3.93e − 1	0.0665
	5.0	14.939	5.60e − 1	0.0361
	6.0	22.013	7.44e − 1	0.0327
0.05	1.0	2.2087	1.87e − 2	0.00840
	2.0	3.3449	6.34e − 2	0.0186
	3.0	5.7845	1.25e − 1	0.0212
	4.0	9.7061	1.99e − 1	0.0337
	5.0	15.214	2.84e − 1	0.0183
	6.0	22.381	3.76e − 1	0.0165

```
% first line of the form
%    function ans=deriv(t,y)
% A sample call would be
%    [t,z]=euler_for(t0,z0,b,delta,'deriv')
%
% Output:
% The routine eulercls will return two vectors, t and y.
% The vector t will contain the node points
%    t(1)=t0, t(j)=t0+(j-1)*h, j=1,2,...,N
% with
%    t(N) <= t_end-h,  t(N)+h > t_end-h
% The vector y will contain the estimates of the solution Y
% at the node points in t.
%
n = fix((t_end-t0)/h)+1;
t = linspace(t0,t0+(n-1)*h,n)';
y = zeros(n,1);
```

```
y(1) = y0;
for i = 2:n
  y(i) = y(i-1)+h*feval(fcn,t(i-1),y(i-1));
end
```

2.2 ERROR ANALYSIS OF EULER'S METHOD

The purpose of analyzing Euler's method is to understand how it works, be able to predict the error when using it, and perhaps accelerate its convergence. Being able to do this for Euler's method will also make it easier to answer the same questions for other, more efficient numerical methods.

For the error analysis, we assume that the initial value problem (1.7) has a unique solution $Y(t)$ on $t_0 \leq t \leq b$, and further, that this solution has a bounded second derivative $Y''(t)$ over this interval. We begin by applying Taylor's theorem to approximating $Y(t_{n+1})$,

$$Y(t_{n+1}) = Y(t_n) + hY'(t_n) + \tfrac{1}{2}h^2 Y''(\xi_n)$$

for some $t_n \leq \xi_n \leq t_{n+1}$. Using the fact that $Y(t)$ satisfies the differential equation,

$$Y'(t) = f(t, Y(t)),$$

our Taylor approximation becomes

$$Y(t_{n+1}) = Y(t_n) + hf(t_n, Y(t_n)) + \tfrac{1}{2}h^2 Y''(\xi_n). \tag{2.9}$$

The term

$$T_{n+1} = \tfrac{1}{2}h^2 Y''(\xi_n) \tag{2.10}$$

is called the *truncation error* for Euler's method, and it is the error in the approximation

$$Y(t_{n+1}) \approx Y(t_n) + hf(t_n, Y(t_n)).$$

To analyze the error in Euler's method, subtract

$$y_{n+1} = y_n + hf(t_n, y_n) \tag{2.11}$$

from (2.9), obtaining

$$\begin{aligned} Y(t_{n+1}) - y_{n+1} = Y(t_n) - y_n + h[f(t_n, Y(t_n)) - f(t_n, y_n)] \\ + \tfrac{1}{2}h^2 Y''(\xi_n). \end{aligned} \tag{2.12}$$

The error in y_{n+1} consists of two parts: (1) the truncation error T_{n+1}, newly introduced at step t_{n+1}; and (2) the *propagated error*

$$Y(t_n) - y_n + h[f(t_n, Y(t_n)) - f(t_n, y_n)].$$

The propagated error can be simplified by applying the mean value theorem to $f(t, z)$, considering it as a function of z,

$$f(t_n, Y(t_n)) - f(t_n, y_n) = \frac{\partial f(t_n, \zeta_n)}{\partial y}[Y(t_n) - y_n] \qquad (2.13)$$

for some ζ_n between $Y(t_n)$ and y_n. Let $e_k \equiv Y(t_k) - y_k$, $k \geq 0$, and then use (2.13) to rewrite (2.12) as

$$e_{n+1} = \left[1 + h\frac{\partial f(t_n, \zeta_n)}{\partial y}\right]e_n + \tfrac{1}{2}h^2 Y''(\xi_n). \qquad (2.14)$$

These results can be used to give a general error analysis of Euler's method for the initial value problem.

Let us first consider a special case that will yield some intuitive understanding of the error in Euler's method. Consider using Euler's method to solve the problem

$$Y'(t) - 2t, \qquad Y(0) = 0, \qquad (2.15)$$

whose true solution is $Y(t) = t^2$. Then, from the error formula (2.14), we have

$$e_{n+1} = e_n + h^2, \qquad e_0 = 0,$$

where we are assuming the initial value $y_0 = Y(0)$. This leads, by induction, to

$$e_n = nh^2, \qquad n \geq 0.$$

Since $nh = t_n$,

$$e_n = ht_n. \qquad (2.16)$$

For each fixed t_n, the error at t_n is proportional to h. The truncation error is $\mathcal{O}(h^2)$, but the cumulative effect of these errors is a total error proportional to h.

We now turn to a convergence analysis of Euler's method for solving the general initial value problem on a finite interval $[t_0, b]$:

$$\begin{aligned} Y'(t) &= f(t, Y(t)), \qquad t_0 \leq t \leq b, \\ Y(t_0) &= Y_0. \end{aligned} \qquad (2.17)$$

For the complete error analysis, we begin with the following lemma. It is quite useful in the analysis of most numerical methods for solving the initial value problem.

Lemma 2.3 *For any real t,*

$$1 + t \leq e^t,$$

and for any $t \geq -1$, any $m \geq 0$,

$$0 \leq (1 + t)^m \leq e^{mt}. \qquad (2.18)$$

Proof. Using Taylor's theorem yields

$$e^t = 1 + t + \tfrac{1}{2}t^2 e^{\xi}$$

with ξ between 0 and t. Since the remainder is never negative, the first result is proved. Formula (2.18) follows easily. ∎

For this and several of the following chapters, we assume that the derivative function $f(t, y)$ satisfies the following stronger Lipschitz condition: there exists $K \geq 0$ such that

$$|f(t, y_1) - f(t, y_2)| \leq K |y_1 - y_2| \qquad (2.19)$$

for $-\infty < y_1, y_2 < \infty$ and $t_0 \leq t \leq b$. Although stronger than necessary, it simplifies the proofs. In addition, given a function $f(t, y)$ satisfying the weaker condition (1.10) and a solution $Y(t)$ to the initial value problem, the function f can be modified to satisfy (2.19) without changing the solution $Y(t)$ or the essential character of the initial value problem (2.17) and its numerical solution.

Theorem 2.4 *Let $f(t, y)$ be a continuous function for $t_0 \leq t \leq b$ and $-\infty < y < \infty$, and further assume that $f(t, y)$ satisfies the Lipschitz condition (2.19). Assume that the solution $Y(t)$ of (2.17) has a continuous second derivative on $[t_0, b]$. Then the solution $\{y_h(t_n) \mid t_0 \leq t_n \leq b\}$ obtained by Euler's method satisfies*

$$\max_{t_0 \leq t_n \leq b} |Y(t_n) - y_h(t_n)| \leq e^{(b-t_0)K} |e_0| + \left[\frac{e^{(b-t_0)K} - 1}{K}\right] \tau(h), \qquad (2.20)$$

where

$$\tau(h) = \tfrac{1}{2} h \, \|Y''\|_\infty = \tfrac{1}{2} h \max_{t_0 \leq t \leq b} |Y''(t)| \qquad (2.21)$$

and $e_0 = Y_0 - y_h(t_0)$.
If, in addition, we have

$$|Y_0 - y_h(t_0)| \leq c_1 h \quad \text{as } h \to 0 \qquad (2.22)$$

for some $c_1 \geq 0$ (e.g., if $Y_0 = y_0$ for all h, then $c_1 = 0$), then there is a constant $B \geq 0$ for which

$$\max_{t_0 \leq t_n \leq b} |Y(t_n) - y_h(t_n)| \leq Bh. \qquad (2.23)$$

Let $e_n = Y(t_n) - y(t_n)$, $n \geq 0$. Let $N \equiv N(h)$ be the integer for which

$$t_N \leq b, \qquad t_{N+1} > b.$$

Define

$$\tau_n = \tfrac{1}{2} h Y''(\xi_n), \qquad 0 \leq n \leq N(h) - 1,$$

based on the truncation error in (2.10). Easily, we obtain

$$\max_{0 \leq n \leq N-1} |\tau_n| \leq \tau(h)$$

using (2.21).
Recalling (2.12), we have

$$e_{n+1} = e_n + h \left[f(t_n, Y_n) - f(t_n, y_n)\right] + h\tau_n. \qquad (2.24)$$

We are using the common notation $Y_n \equiv Y(t_n)$. Taking bounds using (2.19), we obtain

$$|e_{n+1}| \le |e_n| + hK |Y_n - y_n| + h |\tau_n|,$$

$$|e_{n+1}| \le (1 + hK) |e_n| + h\tau(h), \quad 0 \le n \le N(h) - 1. \tag{2.25}$$

Apply this recursively to obtain

$$|e_n| \le (1 + hK)^n |e_0| + \left[1 + (1 + hK) + \cdots + (1 + hK)^{n-1}\right] h\tau(h).$$

Using the formula for the sum of a finite geometric series,

$$1 + r + r^2 + \cdots + r^{n-1} = \frac{r^n - 1}{r - 1}, \quad r \ne 1, \tag{2.26}$$

we obtain

$$|e_n| \le (1 + hK)^n |e_0| + \left[\frac{(1 + hK)^n - 1}{K}\right] \tau(h). \tag{2.27}$$

Using Lemma 2.3, we obtain

$$(1 + hK)^n \le e^{nhK} = e^{(t_n - t_0)K} \le e^{(b - t_0)K},$$

and this with (2.27) implies the main result (2.20).

The remaining result (2.23) is a trivial corollary of (2.20) with the constant B given by

$$B = c_1 e^{(b - t_0)K} + \frac{1}{2} \left[\frac{e^{(b - t_0)K} - 1}{K}\right] \|Y''\|_\infty. \qquad \blacksquare$$

The result (2.23) is consistent with the behavior observed in Tables 2.1 and 2.2 earlier in this chapter, and it agrees with (2.16) for the special case (2.15). When h is halved, the bound Bh is also halved, and that is the behavior in the error observed earlier. Euler's method is said to converge with order 1, because that is the power of h that occurs in the error bound. In general, if we have

$$|Y(t_n) - y_h(t_n)| \le ch^p, \quad t_0 \le t_n \le b \tag{2.28}$$

for some constant $p \ge 0$, then we say that the numerical method is *convergent with order p*. Naturally, the higher the order p, the faster the convergence we can expect.

We emphasize that for the error bound (2.20) to hold, the true solution must be assumed to have a continuous second derivative $Y''(t)$ over $[t_0, b]$. This assumption is not always valid. When $Y''(t)$ does not have such a continuous second derivative, the error bound (2.20) no longer holds. (See Problem 11.)

The error bound (2.20) is valid for a large family of the initial value problems. However, it usually produces a very pessimistic numerical bound for the error, due to the presence of the exponential terms. Under certain circumstances, we can improve the result. Assume

$$\frac{\partial f(t, y)}{\partial y} \le 0, \tag{2.29}$$

$$K \equiv \sup_{\substack{t_0 \le t \le b \\ -\infty < y < \infty}} \left| \frac{\partial f(t,y)}{\partial y} \right| < \infty. \tag{2.30}$$

Note the relation of (2.29) to the stability condition (1.20) in Chapter 1. Also assume that h has been chosen so small that

$$1 - hK \ge -1, \qquad t_0 \le t \le b, \quad -\infty < z < \infty.$$

Returning to (2.14), we have

$$e_{n+1} = e_n + h \frac{\partial f(t_n, \zeta_n)}{\partial y} e_n + \tfrac{1}{2} h^2 Y''(\xi_n) \tag{2.31}$$

with ζ_n between $Y(t_n)$ and y_n. Using (2.29) and (2.30), we have

$$1 \ge 1 + h \frac{\partial f(t_n, \zeta_n)}{\partial y} \ge 1 - hK \ge -1.$$

When combined with (2.31), we have

$$|e_{n+1}| \le |e_n| + ch^2, \qquad t_0 \le t_n \le b, \tag{2.32}$$

where

$$c = \tfrac{1}{2} \|Y''\|_\infty = \tfrac{1}{2} \cdot \max_{t_0 \le t \le b} |Y''(t)|.$$

In addition, assume $e_0 = 0$. Applying (2.32) inductively, we obtain

$$|e_n| \le nch^2 = c(t_n - t_0)h. \tag{2.33}$$

The error is bounded by a quantity proportional to h, and the coefficient of the h term increases linearly with respect to the point t_n, in contrast to the exponential growth given in the bound (2.20).

The error bound in Theorem 2.4 is rigorous, and is useful in providing an insight to the convergence behavior of the numerical solution. However, it is rarely advisable to use (2.20) for an actual error bound, as the next example shows.

Example 2.5 The problem

$$Y'(t) = -Y(t), \qquad Y(0) = 1 \tag{2.34}$$

was solved earlier in this chapter, with the results given in Table 2.1. To apply (2.20), we have $\partial f(t,y)/\partial y = -1$, $K = 1$. The true solution is $Y(t) = e^{-t}$; thus

$$\max_{0 \le t \le b} |Y''(t)| = 1.$$

With $y_0 = Y_0 = 1$, the bound (2.20) becomes

$$\left| e^{-t_n} - y_h(t_n) \right| \le \tfrac{1}{2} h \left(e^b - 1 \right), \qquad 0 \le t_n \le b. \tag{2.35}$$

As $h \to 0$, this shows that $y_h(t)$ converges to e^{-t}. However, this bound is excessively conservative. As b increases, the bound increases exponentially. For $b = 5$, the bound is

$$\left| e^{-t_n} - y_h(t_n) \right| \leq \tfrac{1}{2} h \left(e^5 - 1 \right) \approx 73.7 h, \qquad 0 \leq t_n \leq 5.$$

And this is far larger than the actual errors shown in Table 2.1, by several orders of magnitude. For the problem (2.34), the improved error bound (2.33) applies with $c = \tfrac{1}{2}$ (see Problem 7). A more general approach for accurate error estimation is discussed in the following section. ∎

2.3 ASYMPTOTIC ERROR ANALYSIS

To obtain more accurate predictions of the error, we consider asymptotic error estimates. Assume that Y is 3 times continuously differentiable and

$$\frac{\partial f(t, y)}{\partial y}, \qquad \frac{\partial^2 f(t, y)}{\partial y^2}$$

are both continuous for all values of (t, y) near $(t, Y(t))$, $t_0 \leq t \leq b$. Then one can prove that the error in Euler's method satisfies

$$Y(t_n) - y_h(t_n) = h D(t_n) + \mathcal{O}(h^2), \qquad t_0 \leq t_n \leq b. \tag{2.36}$$

The term $\mathcal{O}(h^2)$ denotes a quantity of maximal size proportional to h^2 over the interval $[t_0, b]$. More generally, the statement

$$F(h; t_n) = \mathcal{O}(h^p), \qquad t_0 \leq t_n \leq b$$

for some constant p means

$$\max_{t_0 \leq t_n \leq b} |F(h; t_n)| \leq c h^p$$

for some constant c and all sufficiently small values of h.

Assuming $y_0 = Y_0$, the usual case, the function $D(t)$ satisfies an initial value problem for a linear differential equation,

$$D'(t) = g(t) D(t) + \tfrac{1}{2} Y''(t), \qquad D(t_0) = 0, \tag{2.37}$$

where

$$g(t) = \frac{\partial f(t, y)}{\partial y} \bigg|_{y = Y(t)}.$$

When $D(t)$ can be obtained explicitly, the leading error term $h D(t_n)$ from the formula (2.36) usually provides a quite good estimate of the true error $Y(t_n) - y_h(t_n)$, and the quality of the estimation improves with decreasing stepsize h.

Example 2.6 Consider again the problem (2.34). Then $D(t)$ satisfies

$$D'(t) = -D(t) + \tfrac{1}{2}e^{-t}, \qquad D(0) = 0.$$

The solution is

$$D(t) = \tfrac{1}{2}te^{-t}.$$

Using (2.36), the error satisfies

$$Y(t_n) - y_h(t_n) \approx \tfrac{1}{2}ht_ne^{-t_n}. \tag{2.38}$$

We are neglecting the $\mathcal{O}(h^2)$ term, since it should be substantially smaller than the term $hD(t)$ in (2.36), for all sufficiently small values of h. To check the accuracy of (2.38), consider $t_n = 5.0$ with $h = 0.05$. Then

$$\tfrac{1}{2}ht_ne^{-t_n} \doteq 0.000842.$$

From Table 2.1, the actual error is 0.000817, which is quite close to our estimate of it. ∎

How do we obtain the result given in (2.36)? We sketch the main ideas but do not fill in all of the details. We begin by approximating the error equation (2.31) with

$$\widehat{e}_{n+1} = \left[1 + h\frac{\partial f(t, Y(t_n))}{\partial y}\right]\widehat{e}_n + \tfrac{1}{2}h^2 Y''(t_n). \tag{2.39}$$

We have used

$$\frac{\partial f(t_n, \zeta_n)}{\partial y} \approx \frac{\partial f(t, Y(t_n))}{\partial y},$$
$$Y''(\xi_n) \approx Y''(t_n).$$

This will cause an approximation error

$$e_n - \widehat{e}_n = \mathcal{O}(h^2), \tag{2.40}$$

although that may not be immediately evident. In addition, we may write

$$\widehat{e}_n = h\delta_n, \qquad n = 0, 1, \ldots, \tag{2.41}$$

on the basis of (2.33); and for simplicity, assume $\delta_0 = 0$.

Substituting (2.41) into (2.39) and then canceling h, we obtain

$$\begin{aligned}
\delta_{n+1} &= \left[1 + h\frac{\partial f(t, Y(t_n))}{\partial y}\right]\delta_n + \tfrac{1}{2}hY''(t_n) \\
&= \delta_n + h\left[\frac{\partial f(t, Y(t_n))}{\partial y}\delta_n + \tfrac{1}{2}Y''(t_n)\right].
\end{aligned}$$

This is Euler's method applied to (2.37). Applying the earlier convergence analysis for Euler's method, we have

$$\max_{t_0 \le t_n \le b} |D(t_n) - \delta_n| \le Bh$$

for some constant $B > 0$. We then multiply by h to get

$$\max_{t_0 \le t_n \le b} |hD(t_n) - \widehat{e}_n| \le Bh^2.$$

Combining this with (2.40) demonstrates (2.36), although we have omitted a number of details.

We comment that the function $D(t)$ defined by (2.37) is continuously differentiable. Then the error formula (2.36) allows us to use the divided difference

$$\frac{y_h(t_{n+1}) - y_h(t_n)}{h}$$

as an approximation to the derivative $Y'(t_n)$ (or $Y'(t_{n+1})$),

$$Y'(t_n) - \frac{y_h(t_{n+1}) - y_h(t_n)}{h} = \mathcal{O}(h). \tag{2.42}$$

The proof of this is left as Problem 16.

2.3.1 Richardson extrapolation

It is not practical to try to find the function $D(t)$ from the problem (2.37), principally because it requires knowledge of the true solution $Y(t)$. The real power of the formula (2.36) is that it describes precisely the error behavior. We can use (2.36) to estimate the solution error and to improve the quality of the numerical solution, without an explicit knowledge of the function $D(t)$. For this purpose, we need two numerical solutions, say, $y_h(t)$ and $y_{2h}(t)$ over the interval $t_0 \le t \le b$.

Assume that t is a node point with the stepsize $2h$, and note that it is then also a node point with the stepsize h. By the formula (2.36), we have

$$Y(t) - y_h(t) = hD(t) + \mathcal{O}(h^2),$$
$$Y(t) - y_{2h}(t) = 2hD(t) + \mathcal{O}(h^2).$$

Multiply the first equation by 2, and then subtract the second equation to eliminate $D(t)$, obtaining

$$Y(t) - [2\,y_h(t) - y_{2h}(t)] = \mathcal{O}(h^2). \tag{2.43}$$

This can also be written as

$$Y(t) - y_h(t) = y_h(t) - y_{2h}(t) + \mathcal{O}(h^2). \tag{2.44}$$

We know from our earlier error analysis that $Y(t) - y_h(t) = \mathcal{O}(h)$. By dropping the higher-order term $\mathcal{O}(h^2)$ in (2.43), we obtain *Richardson's extrapolation formula*

$$Y(t) \approx \widetilde{y}_h(t) \equiv 2y_h(t) - y_{2h}(t). \tag{2.45}$$

Table 2.3 Euler's method with Richardson extrapolation

t	$Y(t) - y_h(t)$	$y_h(t) - y_{2h}(t)$	$\widetilde{y}_h(t)$	$Y(t) - \widetilde{y}_h(t)$
1.0	9.39e − 3	9.81e − 3	3.6829346e − 1	−4.14e − 4
2.0	6.82e − 3	6.94e − 3	1.3544764e − 1	−1.12e − 4
3.0	3.72e − 3	3.68e − 3	4.9748443e − 2	3.86e − 5
4.0	1.80e − 3	1.73e − 3	1.8249877e − 2	6.58e − 5
5.0	8.17e − 4	7.67e − 4	6.6872853e − 3	5.07e − 5

Dropping the higher-order term in (2.44), we obtain *Richardson's error estimate*

$$Y(t) - y_h(t) \approx y_h(t) - y_{2h}(t). \tag{2.46}$$

With these formulas, we can estimate the error in Euler's method and can also obtain a more rapidly convergent solution $\widetilde{y}_h(t)$.

Example 2.7 Consider (2.34) with stepsize $h = 0.05$, $2h = 0.1$. Then Table 2.3 contains Richardson's extrapolation results for selected values of t. Note that (2.46) is a fairly accurate estimator of the error, and that $\widetilde{y}_h(t)$ is much more accurate than $y_h(t)$. ■

Using (2.43), we have

$$Y(t_n) - \widetilde{y}_h(t_n) = \mathcal{O}(h^2), \tag{2.47}$$

an improvement on the convergence order of Euler's method. We will consider again this type of extrapolation for the methods introduced in later chapters. However, the actual formulas may be different from (2.45) and (2.46), and they will depend on the order of the method.

2.4 NUMERICAL STABILITY

Recall the discussion of stability for the initial value problem given in Section 1.2. In particular, recall the result (1.12) bounding the change in the solution $Y(t)$ when the initial condition is perturbed by ε. To perform a similar analysis for Euler's method, we define a numerical solution $\{z_n\}$ by

$$z_{n+1} = z_n + hf(t_n, z_n), \qquad n = 0, 1, \ldots, N(h) - 1 \tag{2.48}$$

with $z_0 = y_0 + \epsilon$. This is analogous to looking at the solution $Y(t; \varepsilon)$ to the perturbed initial value problem, in (1.11). We compare the two numerical solutions $\{z_n\}$ and $\{y_n\}$ as $h \to 0$.

Let $e_n = z_n - y_n$, $n \geq 0$. Then $e_0 = \epsilon$, and subtracting $y_{n+1} = y_n + hf(t_n, y_n)$ from (2.48), we obtain

$$e_{n+1} = e_n + h\left[f(t_n, z_n) - f(t_n, y_n)\right].$$

This has exactly the same form as (2.24), with τ_n set to zero. Using the same procedure as that following (2.24), we have

$$\max_{0 \leq n \leq N(h)} |z_n - y_n| \leq e^{(b-t_0)K} |\epsilon|.$$

Consequently, there is a constant $\widehat{c} \geq 0$, independent of h, such that

$$\max_{0 \leq n \leq N(h)} |z_n - y_n| \leq \widehat{c}|\epsilon|. \tag{2.49}$$

This is the analog to the result (1.12) for the original initial value problem. This says that Euler's method is a stable numerical method for the solution of the initial value problem (2.17). We insist that all numerical methods for initial value problems possess this form of stability, imitating the stability of the original problem (2.17). In addition, we require other forms of stability, based on replicating additional properties of the initial value problem; these are introduced later.

2.4.1 Rounding error accumulation

The finite precision of computer arithmetic affects the accuracy in the numerical solution of a differential equation. To investigate this effect, consider Euler's method (2.5). The simple arithmetic operations and the evaluation of $f(x_n, y_n)$ will usually contain errors due to rounding or chopping. For definitions of chopped and rounded floating-point arithmetic, see [12, p. 39]. Thus what is actually evaluated is

$$\widehat{y}_{n+1} = \widehat{y}_n + hf(x_n, \widehat{y}_n) + \delta_n, \qquad n \geq 0, \qquad \widehat{y}_0 = Y_0. \tag{2.50}$$

The quantity δ_n will be based on the precision of the arithmetic, and its size is affected by that of $\widehat{y}_n$. To simplify our work, we assume simply

$$|\delta_n| \leq cu \cdot \max_{x_0 \leq x \leq x_n} |Y(x)|, \tag{2.51}$$

where u is the *machine epsilon* of the computer (see [12, p. 38]) and c is a constant of magnitude 1 or larger. Using double precision arithmetic with a processor based on the IEEE floating-point arithmetic standard, $u \doteq 2.2 \times 10^{-16}$.

To compare $\{\widehat{y}_n\}$ to the true solution $Y(x)$, we begin by writing

$$Y(x_{n+1}) = Y(x_n) + hf(x_n, Y(x_n)) + \tfrac{1}{2}h^2 Y''(\xi_n), \tag{2.52}$$

which was obtained earlier in (2.9). Subtracting (2.50) from (2.52), we get

$$\begin{aligned} Y(x_{n+1}) - \widehat{y}_{n+1} &= Y(x_n) - \widehat{y}_n + h[f(x_n, Y(x_n)) - f(x_n, \widehat{y}_n)] \\ &\quad + \tfrac{1}{2}h^2 Y''(x_n) - \delta_n, \qquad n \geq 0 \end{aligned} \tag{2.53}$$

with $Y(x_0) - \hat{y}_0 = 0$. This equation is analogous to the error equation given earlier in (2.12), with the role of the truncation error $\frac{1}{2}h^2 Y''(\xi_n)$ in that earlier equation replaced by the term

$$\frac{1}{2}h^2 Y''(\xi_n) - \delta_n = h\left[\frac{1}{2}hY''(\xi_n) - \frac{\delta_n}{h}\right]. \tag{2.54}$$

If the argument in the proof of Theorem 2.4 is applied to (2.53) rather than to (2.12), then the error result (2.20) generalizes to

$$|Y(x_n) - \hat{y}_n| \leq c_1 \left\{\frac{1}{2}h\left[\max_{x_0 \leq x \leq b} |Y''(x)|\right] + \frac{cu}{h}\left[\max_{x_0 \leq x \leq b} |Y(x)|\right]\right\} \tag{2.55}$$

for $x_0 \leq x_n \leq b$, we obtain

$$c_1 = \frac{e^{(b-x_0)K} - 1}{2K},$$

and K is the supremum of $|\partial f(x, y)/\partial y|$, defined in (2.30). The term in braces on the right side of (2.55) is obtained by bounding the term in brackets on the right side of (2.54) and using the assumption (2.51).

In essence, (2.55) says that

$$|Y(x_n) - \hat{y}_n| \leq \alpha_1 h + \frac{\alpha_2}{h}, \qquad x_0 \leq x_n \leq b$$

for appropriate choices of α_1, α_2. Note that α_2 is generally small because u is small. Thus the error bound will initially decrease as h decreases; but at a critical value of h, call it h^*, the error bound will increase, because of the term α_2/h. The same qualitative behavior turns out to apply also for the actual error $Y(x_n) - y_n$. Thus there is a limit on the attainable accuracy, and it is less than the number of digits available in the machine floating-point representation. This same analysis is valid for other numerical methods, with a term of the form

$$\frac{cu}{h}\left[\max_{x_0 \leq x \leq b} |Y(x)|\right]$$

to be included as part of the global error for the numerical method. With rounded floating-point arithmetic, this behavior can usually be improved on. But with chopped floating-point arithmetic, it is likely to be accurate in a qualitative sense: as h is halved, the contribution to the error due to the chopped arithmetic will double.

Example 2.8 Solve the problem

$$Y'(x) = -Y(x) + 2\cos(x), \qquad Y(0) = 1$$

using Euler's method. The true solution is $Y(x) = \sin x + \cos x$. Use a four digit decimal machine with chopped floating-point arithmetic, and then repeat the calculation with rounded floating-point arithmetic. The machine epsilon in this arithmetic is $u = 0.001$. Finally, give the results of Euler's method with exact arithmetic. The

Table 2.4   Effects of rounding/chopping errors in Euler's method

h	x	Chopped arithmetic $Y(x) - \hat{y}_h(x)$	Rounded arithmetic $Y(x) - \hat{y}_h(x)$	Exact arithmetic $Y(x) - y_h(x)$
0.04	1	$-1.00e - 2$	$-1.70e - 2$	$-1.70e - 2$
	2	$-1.17e - 2$	$-1.83e - 2$	$-1.83e - 2$
	3	$-1.20e - 3$	$-2.80e - 3$	$-2.78e - 3$
	4	$1.00e - 2$	$1.60e - 2$	$1.53e - 2$
	5	$1.13e - 2$	$1.96e - 2$	$1.94e - 2$
0.02	1	$7.00e - 3$	$-9.00e - 3$	$-8.46e - 3$
	2	$4.00e - 3$	$-9.10e - 3$	$-9.13e - 3$
	3	$2.30e - 3$	$-1.40e - 3$	$-1.40e - 3$
	4	$-6.00e - 3$	$8.00e - 3$	$7.62e - 3$
	5	$-6.00e - 3$	$8.50e - 3$	$9.03e - 3$
0.01	1	$2.80e - 2$	$-3.00e - 3$	$-4.22e - 3$
	2	$2.28e - 2$	$-4.30e - 3$	$-4.56e - 3$
	3	$7.40e - 3$	$-4.00e - 4$	$-7.03e - 4$
	4	$-2.30e - 2$	$3.00e - 3$	$3.80e - 3$
	5	$-2.41e - 2$	$4.60e - 3$	$4.81e - 3$

results with decreasing h are given in Table 2.4. The errors for the answers that are obtained by using floating–point chopped and/or rounded decimal arithmetic are based on the true answers rounded to four digits.

Note that the errors with the chopped case are affected at $h = 0.02$, with the error at $x = 3$ larger than when $h = 0.04$ for that case. The increasing error is clear with the $h = 0.01$ case, at all points. In contrast, the errors using rounded arithmetic continue to decrease, although the $h = 0.01$ case is affected slightly, in comparison to the true errors when no rounding is present. The column with the errors for the case with exact arithmetic show that the use of the rounded decimal arithmetic has less effect on the error than does the use of chopped arithmetic. But there is still an effect.   ■

PROBLEMS

1. Solve the following problems using Euler's method with stepsizes of $h = 0.2, 0.1, 0.05$. Compute the error and relative error using the true solution $Y(t)$. For selected values of t, observe the ratio by which the error decreases when h is halved.

   (a) $Y'(t) = [\cos(Y(t))]^2$, $0 \le t \le 10$, $Y(0) = 0$;

$Y(t) = \tan^{-1}(t)$.

(b) $Y'(t) = \dfrac{1}{1+t^2} - 2[Y(t)]^2$, $0 \le t \le 10$, $Y(0) = 0$;

$Y(t) = \dfrac{t}{1+t^2}$.

(c) $Y'(t) = \dfrac{1}{4}Y(t)\left[1 - \dfrac{1}{20}Y(t)\right]$, $0 \le t \le 20$, $Y(0) = 1$;

$Y(t) = \dfrac{20}{1 + 19e^{-t/4}}$.

(d) $Y'(t) = -[Y(t)]^2$, $1 \le t \le 10$, $Y(1) = 1$;

$Y(t) = \dfrac{1}{t}$.

(e) $Y'(t) = te^{-t} - Y(t)$, $0 \le t \le 10$, $Y(0) = 1$;

$Y(t) = \left(1 + \dfrac{1}{2}t^2\right)e^{-t}$.

(f) $Y'(t) = \dfrac{t^3}{Y(t)}$, $0 \le t \le 10$, $Y(0) = 1$;

$Y(t) = \sqrt{\dfrac{1}{2}t^4 + 1}$.

(g) $Y'(t) = (3t^2 + 1)Y(t)^2$, $0 \le t \le 10$, $Y(0) = -1$;

$Y(t) = -\left(t^3 + t + 1\right)^{-1}$.

2. Compute the true solution to the problem

$$Y'(t) = -e^{-t}Y(t), \qquad Y(0) = 1.$$

Using Euler's method, solve this equation numerically with stepsizes of $h = 0.2, 0.1, 0.05$. Compute the error and relative error using the true solution $Y(t)$.

3. Consider the linear problem

$$Y'(t) = \lambda Y(t) + (1 - \lambda)\cos(t) - (1 + \lambda)\sin(t), \qquad Y(0) = 1.$$

The true solution is $Y(t) = \sin(t) + \cos(t)$. Solve this problem using Euler's method with several values of λ and h, for $0 \le t \le 10$. Comment on the results.

(a) $\lambda = -1$; $h = 0.5, 0.25, 0.125$.

(b) $\lambda = 1$; $h = 0.5, 0.25, 0.125$.

(c) $\lambda = -5$; $h = 0.5, 0.25, 0.125, 0.0625$.

(d) $\lambda = 5$; $h = 0.125, 0.0625$.

4. As a special case in which the error of Euler's method can be analyzed directly, consider Euler's method applied to

$$Y'(t) = Y(t), \qquad Y(0) = 1.$$

The true solution is e^t.

(a) Show that the solution of Euler's method can be written as

$$y_h(t_n) = (1 + h)^{t_n/h}, \qquad n \geq 0.$$

(b) Using L'Hospital's rule from calculus, show that

$$\lim_{h \to 0} (1 + h)^{1/h} = e.$$

This then proves that for fixed $t = t_n$,

$$\lim_{h \to 0} y_h(t) = e^t.$$

(c) Let us do a more delicate convergence analysis. Use the property $a^b = e^{b \log a}$ to write

$$y_h(t_n) = e^{t_n \log(1+h)/h}.$$

Then use the formula

$$\log(1 + h) = h - \tfrac{1}{2}h^2 + \mathcal{O}(h^3)$$

and Taylor expansion of the natural exponential function to show that

$$Y(t_n) - y_h(t_n) = \tfrac{1}{2}h t_n e^{t_n} + \mathcal{O}(h^2).$$

This shows that for h small, the error is almost proportional to h, a phenomenon already observed from the numerical results given in Tables 2.1 and 2.2.

5. Repeat the general procedures of Problem 4, but do so for the initial value problem

$$Y'(t) = cY(t), \qquad Y(0) = 1$$

with $c \neq 0$ a given constant.

6. Check the accuracy of the error bound (2.35) for $b = 1, 2, 3, 4, 5$ and $h = 0.2, 0.1, 0.05$. Compute the error bound and compare it with Table 2.1.

7. Consider again the problem (2.34) of Example 2.5. Let us derive a more accurate error bound than the one given in Theorem 2.4. From (2.14) we have

$$e_{n+1} = (1 - h) e_n + \tfrac{1}{2}h^2 e^{-\xi_n}.$$

Using this formula with $0 < h \leq 1$, and recalling $e_0 = 0$, show the error bound

$$|e_n| \leq \tfrac{1}{2} h t_n.$$

Compare this error bound to the true errors in Table 2.1.
Hint: $1 - h \leq 1$ and $e^{-\xi_n} \leq 1$.

8. Compute the error bound (2.20), assuming $y_0 = Y_0$, for the problem (2.8) given earlier in this chapter. Compare the bound with the actual errors given in Table 2.2, for $b = 1, 2, 3, 4, 5$ and $h = 0.2, 0.1, 0.05$.

9. Repeat Problem 8 for the equation in Problem 1 (a).

10. For Problems 1 (b)–(d), the constant K in (2.19) will be infinite. To use the error bound (2.20) in such cases, let

$$K = 2 \cdot \max_{t_0 \leq t \leq b} \left| \frac{\partial f(t, Y(t))}{\partial y} \right|.$$

This can be shown to be adequate for all sufficiently small values of h. Then repeat Problem 8 for Problem 1 (b)–(d).

11. Consider the initial value problem

$$Y'(t) = \alpha t^{\alpha - 1}, \quad Y(0) = 0,$$

where $\alpha > 0$. The true solution is $Y(t) = t^\alpha$. When $\alpha \neq$ integer, the true solution is not infinitely differentiable. In particular, to have Y twice continuously differentiable, we need $\alpha \geq 2$. Use the Euler method to solve the initial value problem for $\alpha = 2.5, 1.5, 1.1$ with stepsize $h = 0.2, 0.1, 0.05$. Compute the solution errors at the nodes, and determine numerically the convergence orders of the Euler method for these problems.

12. The solution of

$$Y'(t) = \lambda Y(t) + \cos(t) - \lambda \sin(t), \quad Y(0) = 0$$

is $Y(t) = \sin(t)$. Find the asymptotic error formula (2.36) in this case. Also compute the Euler solution for $0 \leq t \leq 6$, $h = 0.2, 0.1, 0.05$, and $\lambda = 1, -1$. Compare the true errors with those obtained from the asymptotic estimate

$$Y(t_n) - y_n \approx h D(t_n).$$

13. Repeat Problem 12 for Problem 1 (d). Compare for $1 \leq t \leq 6, h = 0.2, 0.1, 0.05$.

14. For the example (2.8), with the numerical results in Table 2.2, use Richardson's extrapolation to estimate the error $Y(t_n) - y_h(t_n)$ when $h = 0.05$. Also, produce the Richardson extrapolate $\widetilde{y}_h(t_n)$ and compute its error. Do this for $t_n = 1, 2, 3, 4, 5, 6$.

15. Repeat Problem 14 for Problems 1 (a)–(d).

16. Use Taylor's theorem to show the standard numerical differentiation method

$$Y'(t_{n+1}) = \frac{Y(t_{n+1}) - Y(t_n)}{h} + \mathcal{O}(h).$$

Combine this with (2.36) to prove the error result (2.42).

CHAPTER 3

SYSTEMS OF DIFFERENTIAL EQUATIONS

Although some applications of differential equations involve only a single first-order equation, most applications involve a system of several such equations or higher-order equations. In this chapter, we consider systems of first-order equations, showing how Euler's method applies to such systems. Numerical treatment of higher-order equations can be carried out by first converting them to equivalent systems of first-order equations.

To begin with a simple case, the general form of a system of two first-order differential equations is

$$
\begin{aligned}
Y_1'(t) &= f_1(t, Y_1(t), Y_2(t)), \\
Y_2'(t) &= f_2(t, Y_1(t), Y_2(t)).
\end{aligned}
\tag{3.1}
$$

The functions $f_1(t, z_1, z_2)$ and $f_2(t, z_1, z_2)$ define the differential equations, and the unknown functions $Y_1(t)$ and $Y_2(t)$ are being sought. The initial value problem consists of solving (3.1), subject to the initial conditions

$$
Y_1(t_0) = Y_{1,0}, \qquad Y_2(t_0) = Y_{2,0}.
\tag{3.2}
$$

Example 3.1

(a) The initial value problem

$$Y_1'(t) = Y_1(t) - 2Y_2(t) + 4\cos(t) - 2\sin(t), \quad Y_1(0) = 1,$$
$$Y_2'(t) = 3Y_1(t) - 4Y_2(t) + 5\cos(t) - 5\sin(t), \quad Y_2(0) = 2 \tag{3.3}$$

has the solution

$$Y_1(t) = \cos(t) + \sin(t), \quad Y_2(t) = 2\cos(t).$$

This example will be used later in a numerical example illustrating Euler's method for systems.

(b) Consider the system

$$Y_1'(t) = AY_1(t)[1 - BY_2(t)], \quad Y_1(0) = Y_{1,0},$$
$$Y_2'(t) = CY_2(t)[DY_1(t) - 1], \quad Y_2(0) = Y_{2,0} \tag{3.4}$$

with constants $A, B, C, D > 0$. This is called the Lotka–Volterra predator–prey model. The variable t denotes time, $Y_1(t)$ the number of prey (e.g., rabbits) at time t, and $Y_2(t)$ the number of predators (e.g., foxes). If there is only a single type of predator and a single type of prey, then this model is often a reasonable approximation of reality. The behavior of the solutions Y_1 and Y_2 is illustrated in Problem 8. ∎

The initial value problem for a system of m first-order differential equations has the general form

$$Y_1'(t) = f_1(t, Y_1(t), \ldots, Y_m(t)), \quad Y_1(t_0) = Y_{1,0},$$
$$\vdots \tag{3.5}$$
$$Y_m'(t) = f_m(t, Y_1(t), \ldots, Y_m(t)), \quad Y_m(t_0) = Y_{m,0}.$$

We seek the functions $Y_1(t), \ldots, Y_m(t)$ on some interval $t_0 \le t \le b$. An example of a three-equation system is given later in (3.21).

The general form (3.5) is clumsy to work with, and it is not a convenient way to specify the system when using a computer program for its solution. To simplify the form of (3.5), represent the solution and the differential equations by using column vectors. Denote

$$\mathbf{Y}(t) = \begin{bmatrix} Y_1(t) \\ \vdots \\ Y_m(t) \end{bmatrix}, \quad \mathbf{Y}_0 = \begin{bmatrix} Y_{1,0} \\ \vdots \\ Y_{m,0} \end{bmatrix}, \quad \mathbf{f}(t, \mathbf{y}) = \begin{bmatrix} f_1(t, y_1, \ldots, y_m) \\ \vdots \\ f_m(t, y_1, \ldots, y_m) \end{bmatrix} \tag{3.6}$$

with $\mathbf{y} = [y_1, y_2, \ldots, y_m]^T$. Then (3.5) can be rewritten as

$$\mathbf{Y}'(t) = \mathbf{f}(t, \mathbf{Y}(t)), \quad \mathbf{Y}(t_0) = \mathbf{Y}_0. \tag{3.7}$$

This resembles the earlier first-order single equation, but it is general as to the number of equations. Computer programs for solving systems will almost always refer to the system in this manner.

Example 3.2 System (3.3) can be rewritten as

$$\mathbf{Y}'(t) = A\mathbf{Y}(t) + \mathbf{G}(t), \qquad \mathbf{Y}(0) = \mathbf{Y}_0$$

with

$$\mathbf{Y} = \begin{bmatrix} Y_1 \\ Y_2 \end{bmatrix}, \qquad A = \begin{bmatrix} 1 & -2 \\ 3 & -4 \end{bmatrix},$$

$$\mathbf{G}(t) = \begin{bmatrix} 4\cos(t) - 2\sin(t) \\ 5\cos(t) - 5\sin(t) \end{bmatrix}, \qquad \mathbf{Y}_0 = \begin{bmatrix} 1 \\ 2 \end{bmatrix}.$$

In the notation of (3.6), we obtain

$$\mathbf{f}(t, \mathbf{y}) = A\mathbf{y} + \mathbf{G}(t), \qquad \mathbf{y} = [y_1, y_2]^{\mathsf{T}}. \qquad \blacksquare$$

The general theory in Chapter 1 for a single differential equation generalizes in an easy way to systems of first-order differential equations, once we have introduced appropriate notation and tools for (3.6). For example, the role of the partial differential $\partial f / \partial y$ is replaced with the Jacobian matrix

$$\mathbf{f}_{\mathbf{y}}(t, \mathbf{y}) = \left[\frac{\partial f_i(t, y_1, \ldots, y_m)}{\partial y_j} \right]_{i,j=1}^{m}. \tag{3.8}$$

We replace the absolute value $|\cdot|$ with a vector norm. A convenient choice is the *maximum norm*:

$$\|\mathbf{y}\|_{\infty} = \max_{1 \le i \le m} |y_i|, \qquad \mathbf{y} \in \mathbb{R}^m.$$

With this, we can generalize the Lipschitz condition (2.19) to

$$\|\mathbf{f}(t, \mathbf{y}) - \mathbf{f}(t, \mathbf{z})\|_{\infty} \le K \|\mathbf{y} - \mathbf{z}\|_{\infty}, \qquad \mathbf{y}, \mathbf{z} \in \mathbb{R}^m, \quad t_0 \le t \le b, \tag{3.9}$$

$$K = \max_{t_0 \le t \le b} \max_{1 \le i \le m} \sup_{\mathbf{y} \in \mathbb{R}^m} \sum_{j=1}^{m} \left| \frac{\partial f_i(t, \mathbf{y})}{\partial y_j} \right|.$$

3.1 HIGHER-ORDER DIFFERENTIAL EQUATIONS

In physics and engineering, the use of *Newton's second law of motion* leads to systems of second-order differential equations, modeling some of the most important physical phenomena of nature. In addition, other applications also lead to higher-order equations. Higher-order equations can be studied either directly or through equivalent systems of first-order equations.

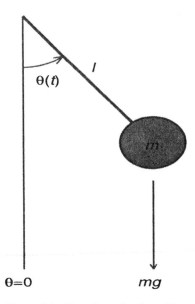

Figure 3.1 The schematic of pendulum

As an example, consider the second-order equation

$$Y''(t) = f(t, Y(t), Y'(t)),$$ (3.10)

where $f(t, y_1, y_2)$ is given. The initial value problem consists of solving (3.10) subject to the initial conditions

$$Y(t_0) = Y_0, \qquad Y'(t_0) = Y_0'.$$ (3.11)

To reformulate this as a system of first-order equations, denote

$$Y_1(t) = Y(t), \qquad Y_2(t) = Y'(t).$$

Then Y_1 and Y_2 satisfy

$$
\begin{aligned}
Y_1'(t) &= Y_2(t), & Y_1(t_0) &= Y_0, \\
Y_2'(t) &= f(t, Y_1(t), Y_2(t)), & Y_2(t_0) &= Y_0'.
\end{aligned}
$$ (3.12)

Also, starting from this system, it is straightforward to show that the solution Y_1 of (3.12) will also have to satisfy (3.10) and (3.11), thus demonstrating the equivalence of the two formulations.

Example 3.3 Consider the pendulum shown in Figure 3.1, of mass m and length l. The motion of this pendulum about its centerline $\theta = 0$ is modeled by a second-order differential equation derived from Newton's second law of motion. If the pendulum is assumed to move back and forth with negligible friction at its vertex, then the motion is modeled fairly accurately by the equation

$$ml\frac{d^2\theta}{dt^2} = -mg\sin(\theta(t)),\tag{3.13}$$

where t is time and $\theta(t)$ is the angle between the vertical centerline and the pendulum. The description of the motion is completed by specifying the initial position $\theta(0)$ and initial angular velocity $\theta'(0)$. To convert this to a system of two first-order equations, we may write

$$Y_1(t) = \theta(t), \qquad Y_2(t) = \theta'(t).$$

Then (3.13) and the initial conditions can be rewritten as

$$\begin{aligned}
Y_1'(t) &= Y_2(t), & Y_1(0) &= \theta(0) \\
Y_2'(t) &= -\frac{g}{l}\sin(Y_1(t)), & Y_2(0) &= \theta'(0).
\end{aligned}\tag{3.14}$$

This system is equivalent to the initial value problem for the original second-order equation (3.13). ■

A general differential equation of order m can be written as

$$\frac{d^m Y(t)}{dt^m} = f\left(t, Y(t), \frac{dY(t)}{dt}, \dots, \frac{d^{m-1}Y(t)}{dt^{m-1}}\right),\tag{3.15}$$

and the initial conditions needed to solve it are given by

$$Y(t_0) = Y_0, \quad Y'(t_0) = Y_0', \quad \dots, \quad Y^{(m-1)}(t_0) = Y_0^{(m-1)}.\tag{3.16}$$

It is reformulated as a system of m first-order equations by introducing

$$Y_1(t) = Y(t), \quad Y_2(t) = Y'(t), \quad \dots, \quad Y_m(t) = Y^{(m-1)}(t).$$

Then the equivalent initial value problem for a system of first-order equations is

$$\begin{aligned}
Y_1'(t) &= Y_2(t), & Y_1(t_0) &= Y_0, \\
&\ \ \vdots & &\ \ \vdots \\
Y_{m-1}'(t) &= Y_m(t), & Y_{m-1}(t_0) &= Y_0^{(m-2)}, \\
Y_m'(t) &= f(t, Y_1(t), \dots, Y_m(t)), & Y_m(t_0) &= Y_0^{(m-1)}.
\end{aligned}\tag{3.17}$$

A special case of (3.15) is the order m linear differential equation

$$\frac{d^m Y}{dt^m} = a_0(t)Y + a_1(t)\frac{dY}{dt} + \dots + a_{m-1}(t)\frac{d^{m-1}Y}{dt^{m-1}} + b(t).\tag{3.18}$$

This is reformulated as above, with

$$Y'_m = a_0(t)Y_1 + a_1(t)Y_2 + \cdots + a_{m-1}(t)Y_m + b(t) \tag{3.19}$$

replacing the last equation in (3.17).

Example 3.4 The initial value problem

$$Y'''(t) + 3Y''(t) + 3Y'(t) + Y(t) = -4\sin(t),$$
$$Y(0) = Y'(0) = 1, \quad Y''(0) = -1 \tag{3.20}$$

is reformulated as

$$
\begin{array}{ll}
Y'_1(t){=}Y_2(t), & Y_1(0){=}1, \\
Y'_2(t){=}Y_3(t), & Y_2(0){=}1, \qquad (3.21) \\
Y'_3(t){=}{-}Y_1(t) - 3Y_2(t) - 3Y_3(t) - 4\sin(t), & Y_3(0){=}{-}1.
\end{array}
$$

The solution of (3.20) is $Y(t) = \cos(t) + \sin(t)$, and the solution of (3.21) can be generated from it. This system will be solved numerically later in this chapter. ∎

3.2 NUMERICAL METHODS FOR SYSTEMS

Euler's method and the numerical methods discussed in later chapters can be applied without change to the solution of systems of first-order differential equations. The numerical method should be applied to each equation in the system, or more simply, in a straightforward way to the system written in the matrix–vector format (3.7). The derivation of numerical methods for the solution of systems is essentially the same as is done for a single equation. The convergence and stability analyses are also done in the same manner.

To be more specific, we consider Euler's method for the general system of two first-order equations that is given in (3.1). By following the derivation given for Euler's method in obtaining (2.9), Taylor's theorem gives

$$Y_1(t_{n+1}) = Y_1(t_n) + h f_1(t_n, Y_1(t_n), Y_2(t_n)) + \frac{h^2}{2} Y_1''(\xi_n),$$

$$Y_2(t_{n+1}) = Y_2(t_n) + h f_2(t_n, Y_1(t_n), Y_2(t_n)) + \frac{h^2}{2} Y_2''(\zeta_n)$$

for some ξ_n, ζ_n in $[t_n, t_{n+1}]$. Dropping the error terms, we obtain Euler's method for a system of two equations for $n \geq 0$:

$$y_{1,n+1} = y_{1,n} + h f_1(t_n, y_{1,n}, y_{2,n}),$$
$$y_{2,n+1} = y_{2,n} + h f_2(t_n, y_{1,n}, y_{2,n}). \tag{3.22}$$

In matrix–vector format, this is

$$\mathbf{y}_{n+1} = \mathbf{y}_n + h\mathbf{f}(t_n, \mathbf{y}_n), \qquad \mathbf{y}_0 = \mathbf{Y}_0. \tag{3.23}$$

The convergence and stability theory of Euler's method and of the other numerical methods also generalizes. The key is to use the matrix–vector notation introduced earlier in the chapter together with (3.8)–(3.9). This allows a straightforward imitation of the proofs given in earlier chapters for a single equation.

Let $m = 2$ as above, and consider Euler's method (3.22) together with the exact initial values $y_{1,0} = Y_{1,0}$, $y_{2,0} = Y_{2,0}$. If $Y_1(t)$, $Y_2(t)$ are twice continuously differentiable, then it can be shown that

$$|Y_1(t_n) - y_{1,n}| \leq ch, \qquad |Y_2(t_n) - y_{2,n}| \leq ch$$

for all $t_0 \leq t_n \leq b$, for some constant c. In addition, the earlier asymptotic error formula (2.36) will still be valid; for $j = 1, 2$, we obtain

$$Y_j(t_n) - y_{j,n} = D_j(t_n)h + \mathcal{O}(h^2), \qquad t_0 \leq t_n \leq b.$$

Thus Richardson's extrapolation and error estimation formulas will still be valid. The functions $D_1(t)$, $D_2(t)$ satisfy a particular linear system of differential equations, but we omit it here. Stability results for Euler's method generalize without any significant change. Thus in summary, the earlier work for Euler's method generalizes without significant change to systems. The same is true of the other numerical methods given earlier, thus justifying our limitation to a single equation for introducing those methods.

MATLAB® program. The following is a MATLAB code `eulersys` implementing the Euler method to solve the initial value problem (3.7). It can be seen that the code `eulersys` is just a slight modification of the code `euler_for` for solving a single equation in Chapter 2. The program can automatically determine the number of equations in the system.

```
function [t,y] = eulersys(t0,y0,t_end,h,fcn)
%
% function [t,y]=eulersys(t0,y0,t_end,h,fcn)
%
% Solve the initial value problem of a system
% of first order equations
%    y' = f(t,y),  t0 <= t <= b,  y(t0)=y0
% Use Euler's method with a stepsize of h.
% The user must supply a program to compute the
% right hand side function with some name, say
% deriv, and a first line of the form
%    function ans=deriv(t,y)
% A sample call would be
%    [t,z]=eulersys(t0,z0,b,delta,'deriv')
```

Table 3.1   Solution of (3.3) using Euler's method

j	t	$Y_j(t)$	$Y_j(t) - y_{j,2h}(t)$	$Y_j(t) - y_{j,h}(t)$	Ratio	$y_{j,h}(t) - y_{j,2h}(t)$
1	2	0.49315	$-5.65e-2$	$-2.82e-2$	2.0	$-2.83e-2$
	4	-1.41045	$-5.64e-3$	$-2.72e-3$	2.1	$-2.92e-3$
	6	0.68075	$4.81e-2$	$2.36e-2$	2.0	$2.44e-2$
	8	0.84386	$-3.60e-2$	$-1.79e-2$	2.0	$-1.83e-2$
	10	-1.38309	$-1.81e-2$	$-8.87e-3$	2.0	$-9.40e-2$
2	2	-0.83229	$-3.36e-2$	$-1.70e-2$	2.0	$-1.66e-2$
	4	-1.30729	$5.94e-3$	$3.19e-3$	1.9	$2.75e-3$
	6	1.92034	$1.59e-2$	$7.69e-3$	2.1	$8.17e-3$
	8	-0.29100	$-2.08e-2$	$-1.05e-2$	2.0	$-1.03e-2$
	10	-1.67814	$1.26e-3$	$9.44e-4$	1.3	$3.11e-4$

```
%
% The program automatically determines the
% number of equations from the dimension of
% the initial value vector y0.
%
% Output:
% The routine eulersys will return a vector t
% and a matrix y. The vector t will contain the
% node points in [t0,t_end]:
%    t(1)=t0, t(j)=t0+(j-1)*h, j=1,2,...,N
% The matrix y is of size N by m, with m the
% number of equations.  The i-th row y(i,:) will
% contain the estimates of the solution Y
% at the node points in t(i).
%
m = length(y0);
n = fix((t_end-t0)/h)+1;
t = linspace(t0,t0+(n-1)*h,n)';
y = zeros(n,m);
y(1,:) = y0;
for i = 2:n
  y(i,:) = y(i-1,:) + h*feval(fcn,t(i-1),y(i-1,:));
end
```

Example 3.5

(a) Solve (3.3) using Euler's method. The numerical results are given in Table 3.1, along with Richardson's error estimate

$$Y_j(t_n) - y_{j,h}(t_n) \approx y_{j,h}(t_n) - y_{j,2h}(t_n), \qquad j = 1, 2.$$

In the table, $h = 0.05$, $2h = 0.1$. It can be seen that this error estimate is quite accurate, except for the one case $j = 2, t = 10$. To get the numerical solution values and their errors at the specified node points $t = 2, 4, 6, 8, 10$, we used the following MATLAB commands, which can be included at the end of the program eulersys for this example.

```
n1 = (n-1)/5;
for i = n1+1:n1:n
e(i,1) = cos(t(i))+sin(t(i))-y(i,1);
e(i,2) = 2*cos(t(i))-y(i,2);
end
diary euler_sys1
fprintf(' h = 6.5f\n', h)
disp(' t y(1) e(1) y(2) e(2)')
for i = n1+1:n1:n
fprintf('2.0f%10.2e%10.2e%10.2e%10.2e\n', ...
t(i), y(i,1),e(i,1),y(i,2),e(i,2))
end
diary off
```

The right-hand side function for this example is defined by the following.

```
function z = eulersys_fcn(t,y);
z = zeros(1,2);
z(1) = y(1)-2*y(2)+4*cos(t)-2*sin(t);
z(2) = 3*y(1)-4*y(2)+5*cos(t)-5*sin(t);
```

(b) Solve the third-order equation in (3.20), using Euler's method to solve the reformulated problem (3.21). The results for $y(t) = Y_1(t) = \sin(t) + \cos(t)$ are given in Table 3.2, for stepsizes $2h = 0.1$ and $h = 0.05$. The Richardson error estimate is again quite accurate. ∎

Other numerical methods apply to systems in the same straightforward manner. Also, by using the matrix form (3.7) for a system, there is no apparent change in the numerical method. For example, the Runge–Kutta method (5.20), given in Section 5.2 of Chapter 5, is

$$y_{n+1} = y_n + \frac{h}{2}[f(t_n, y_n) + f(t_{n+1}, y_n + hf(t_n, y_n))], \qquad n \geq 0. \tag{3.24}$$

Table 3.2 Solution of (3.20) using Euler's method

t	$y(t)$	$y(t) - y_{2h}(t)$	$y(t) - y_h(t)$	Ratio	$y_h(t) - y_{2h}(t)$
2	0.49315	$-8.78e - 2$	$-4.25e - 2$	2.1	$-4.53e - 2$
4	-1.41045	$1.39e - 1$	$6.86e - 2$	2.0	$7.05e - 2$
6	0.68075	$5.19e - 2$	$2.49e - 2$	2.1	$2.70e - 2$
8	0.84386	$-1.56e - 1$	$-7.56e - 2$	2.1	$-7.99e - 2$
10	-1.38309	$8.39e - 2$	$4.14e - 2$	2.0	$4.25e - 2$

Interpret this for a system of two equations with

$$\mathbf{y}_n = \begin{bmatrix} y_{1,n} \\ y_{2,n} \end{bmatrix}, \quad \mathbf{f}(t_n, \mathbf{y}_n) = \begin{bmatrix} f_1(t_n, y_{1,n}, y_{2,n}) \\ f_2(t_n, y_{1,n}, y_{2,n}) \end{bmatrix},$$

$$\mathbf{y}_{n+1} = \mathbf{y}_n + \tfrac{1}{2}h[\mathbf{f}(t_n, \mathbf{y}_n) + \mathbf{f}(t_{n+1}, \mathbf{y}_n + h\mathbf{f}(t_n, \mathbf{y}_n))], \qquad n \geq 0. \qquad (3.25)$$

In component form, the method is

$$y_{j,n+1} = y_{j,n} + \tfrac{1}{2}h[f_j(t_n, y_{1,n}, y_{2,n})$$
$$+ f_j(t_{n+1}, y_{1,n} + hf_1(t_n, y_{1,n}, y_{2,n}), \qquad (3.26)$$
$$y_{2,n} + hf_2(t_n, y_{1,n}, y_{2,n}))]$$

for $j = 1, 2$. The matrix–vector format (3.25) can be programmed very conveniently on a computer. We leave its illustration to the problems.

PROBLEMS

1. Let

$$A = \begin{bmatrix} 1 & -2 \\ 2 & -1 \end{bmatrix}, \qquad \mathbf{Y} = \begin{bmatrix} Y_1 \\ Y_2 \end{bmatrix},$$

$$\mathbf{G}(t) = \begin{bmatrix} -2e^{-t} + 2 \\ -2e^{-t} + 1 \end{bmatrix}, \qquad \mathbf{Y}_0 = \begin{bmatrix} 1 \\ 1 \end{bmatrix}.$$

Write out the two equations that make up the system

$$\mathbf{Y}'(t) = A\mathbf{Y}(t) + \mathbf{G}(t), \qquad \mathbf{Y}(t_0) = \mathbf{Y}_0.$$

The true solution is $\mathbf{Y}(t) = [e^{-t}, 1]^T$.

2. Express the system (3.21) to the general form of Problem 1, giving the matrix A.

3. Convert the following higher-order equations to systems of first-order equations.

(a) $Y'''(t) + 4Y''(t) + 5Y'(t) + 2Y(t) = 2t^2 + 10t + 8$,
$Y(0) = 1, Y'(0) = -1, Y''(0) = 3$.
The true solution is $Y(t) = e^{-t} + t^2$.

(b) $Y''(t) + 4Y'(t) + 13Y(t) = 40\cos(t)$,
$Y(0) = 3, Y'(0) = 4$.
The true solution is $Y(t) = 3\cos(t) + \sin(t) + e^{-2t}\sin(3t)$.

4. Convert the following system of second-order equations to a larger system of first-order equations. This system arises from studying the gravitational attraction of one mass by another:

$$x''(t) = \frac{-cx(t)}{r(t)^3}, \qquad y''(t) = \frac{-cy(t)}{r(t)^3}, \qquad z''(t) = \frac{-cz(t)}{r(t)^3}$$

Here c is a positive constant and $r(t) = [x(t)^2 + y(t)^2 + z(t)^2]^{1/2}$, with t denoting time.

5. Using Euler's method, solve the system in Problem 1. Use stepsizes of $h = 0.1, 0.05, 0.025$, and solve for $0 \le t \le 10$. Use Richardson's error formula to estimate the error for $h = 0.025$.

6. Repeat Problem 5 for the systems in Problem 3.

7. Consider solving the pendulum equation (3.13) with $l = 1$ and $g = 32.2$ ft/s^2. For the initial values, choose $0 < \theta(0) \le \pi/2, \theta'(0) = 0$. Use Euler's method to solve (3.14), and experiment with various values of h so as to obtain a suitably small error in the computed solution. Graph t vs. $\theta(t)$, t vs. $\theta'(t)$, and $\theta(t)$ vs. $\theta'(t)$. Does the motion appear to be periodic in time?

8. Solve the Lotka–Volterra predator–prey model of (3.4) with the parameters $A = 4, B = \frac{1}{2}, C = 3, D = \frac{1}{3}$, and use eulersys to solve approximately this model for $0 \le t \le 5$. Use stepsizes $h = 0.001, 0.0005, 0.00025$. Use the initial values $x(0) = 3, y(0) = 5$. Plot x and y as functions of t, and plot x versus y. Comment on your results. We return to this problem in later chapters when we have more efficient methods for its solution.

CHAPTER 4

THE BACKWARD EULER METHOD AND THE TRAPEZOIDAL METHOD

In Section 1.2 of Chapter 1, we discussed the stability property of the initial value problem (1.7). Roughly speaking, *stability* means that a small perturbation in the initial value of the problem leads to a small change in the solution. In Section 2.4 of Chapter 2, we showed that an analogous stability result was true for Euler's method. In general, we want to work with numerical methods for solving the initial value problem that are numerically stable. This means that for any sufficiently small stepsize h, a small change in the initial value will lead to a small change in the numerical solution. Indeed, such a stability property is closely related to the convergence of the numerical method, a topic we discuss at length in Chapter 7. For another example of the relation between convergence and stability, we refer to Problem 16 for a numerical method that is neither convergent nor stable.

A stable numerical method is one for which the numerical solution is well behaved when considering small perturbations, provided that the stepsize h is sufficiently small. In actual computations, however, the stepsize h cannot be too small since a very small stepsize decreases the efficiency of the numerical method. As can be shown, the accuracy of the forward difference approximations, such as $[Y(t + h) - Y(t)]/h$ to the derivative $Y'(t)$, deteriorates when, roughly speaking, h is of the order of the square root of the *machine epsilon*. Hence, for actual computations, what matters is the performance of the numerical method when h is not assumed *very small*. We need to further analyze the stability of numerical methods when h is not assumed to be small.

Examining the stability question for the general problem

$$Y'(t) = f(t, Y(t)), \qquad Y(t_0) = Y_0 \tag{4.1}$$

is too complicated. Instead, we examine the stability of numerical methods for the *model problem*

$$Y'(t) = \lambda Y(t) + g(t), \qquad Y(0) = Y_0 \tag{4.2}$$

whose exact solution can be found from (1.5). Questions regarding stability and convergence are more easily answered for this problem, and the answers to these questions can be shown to usually be the answers to those same questions for the more general problem (4.1).

Let $Y(t)$ be the solution of (4.2), and let $Y_\epsilon(t)$ be the solution with the perturbed initial data $Y_0 + \epsilon$:

$$Y'_\epsilon(t) = \lambda Y_\epsilon(t) + g(t), \qquad Y_\epsilon(0) = Y_0 + \epsilon.$$

Let $Z_\epsilon(t)$ denote the change in the solution

$$Z_\epsilon(t) = Y_\epsilon(t) - Y(t).$$

Then, subtracting (4.2) from the equation for $Y_\epsilon(t)$, we obtain

$$Z'_\epsilon(t) = \lambda Z_\epsilon(t), \qquad Z_\epsilon(0) = \epsilon.$$

The solution is

$$Z_\epsilon(t) = \epsilon e^{\lambda t}.$$

Typically in applications, we are interested in the case that either λ is real and negative or λ is complex with a negative real part. In such a case, $Z_\epsilon(t)$ will go to zero as $t \to \infty$ and, thus, the effect of the ϵ perturbation dies out for large values of t. (See a related discussion in Section 1.2 of Chapter 1.) We would like the same behavior to hold for the numerical method that is being applied to (4.2).

By considering the function $Z_\epsilon(t)/\epsilon$ instead of $Z_\epsilon(t)$, we obtain the following model problem that is generally used to test the performance of various numerical methods:

$$\begin{aligned} Y' &= \lambda Y, \quad t > 0, \\ Y(0) &= 1. \end{aligned} \tag{4.3}$$

In the following, when we refer to the model problem (4.3), we always assume that the constant $\lambda < 0$ or λ is complex and with $\mathrm{Real}(\lambda) < 0$. The true solution of the problem (4.3) is

$$Y(t) = e^{\lambda t}, \tag{4.4}$$

which decays exponentially in t since the parameter λ has a negative real part.

The kind of stability property we would like for a numerical method is that when it is applied to (4.3), the numerical solution satisfies

$$y_h(t_n) \to 0 \quad \text{as} \quad t_n \to \infty \tag{4.5}$$

for any choice of the stepsize h. The set of values $h\lambda$, considered as a subset of the complex plane, for which $y_n \to 0$ as $n \to \infty$, is called the *region of absolute stability* of the numerical method. The use of $h\lambda$ arises naturally from the numerical method, as we will see.

Let us examine the performance of the Euler method on the model problem (4.3). We have

$$y_{n+1} = y_n + h\lambda\, y_n = (1 + h\lambda)\, y_n, \quad n \ge 0, \quad y_0 = 1.$$

By an inductive argument, it is not difficult to find

$$y_n = (1 + h\lambda)^n, \quad n \ge 0. \tag{4.6}$$

Note that for a fixed node point $t_n = n\,h \equiv \bar{t}$, as $n \to \infty$, we obtain

$$y_n = \left(1 + \frac{\lambda \bar{t}}{n}\right)^n \to e^{\lambda \bar{t}}.$$

The limiting behavior is obtained using L'Hospital's rule from calculus. This confirms the convergence of the Euler method. We emphasize that this is an asymptotic property in the sense that it is valid in the limit as $h \to 0$.

From formula (4.6), we see that $y_n \to 0$ as $n \to \infty$ if and only if

$$|1 + h\lambda| < 1.$$

For λ real and negative, the condition becomes

$$-2 < h\lambda < 0. \tag{4.7}$$

This sets a restriction on the range of h that we can take to apply Euler's method, namely, $0 < h < -2/\lambda$.

Example 4.1 Consider the model problem with $\lambda = -100$. Then the Euler method will perform well only when $h < 2 \times 100^{-1} = 0.02$. The true solution $Y(t) = e^{-100t}$ at $t = 0.2$ is 2.061×10^{-9}. Table 4.1 lists the Euler solution at $t = 0.2$ for several values of h. ∎

4.1 THE BACKWARD EULER METHOD

Now we consider a numerical method that has the property (4.5) for any stepsize h when applied to the model problem (4.3). Such a method is said to be *absolutely stable*.

Table 4.1 Euler's solution at $x = 0.2$ for Example 4.1

h	$y_h(0.2)$
0.1	81
0.05	256
0.02	1
0.01	0
0.001	7.06e $-$ 10

In the derivation of the Euler method, we used the forward difference approximation

$$Y'(t) \approx \frac{1}{h}[Y(t+h) - Y(t)].$$

Let us use, instead, the *backward difference approximation*

$$Y'(t) \approx \frac{1}{h}[Y(t) - Y(t-h)]. \tag{4.8}$$

Then the differential equation $Y'(t) = f(t, Y(t))$ at $t = t_n$ is discretized as

$$y_n = y_{n-1} + h\,f(t_n, y_n).$$

Shifting the index by 1, we then obtain the *backward Euler method*

$$\begin{cases} y_{n+1} = y_n + h\,f(t_{n+1}, y_{n+1}), & 0 \le n \le N-1, \\ y_0 = Y_0. \end{cases} \tag{4.9}$$

Like the Euler method, the backward Euler method is of first-order accuracy, and a convergence result similar to Theorem 2.4 holds. Also, an asymptotic error expansion of the form (2.36) is valid. The method of proof is a variation on that used for Euler's method in Section 2.3 of Chapter 2.

Let us show that the backward Euler method has the desired property (4.5) on the model problem (4.3). We have

$$y_{n+1} = y_n + h\lambda\,y_{n+1},$$
$$y_{n+1} = (1 - h\lambda)^{-1}y_n, \qquad n \ge 0.$$

Using this together with $y_0 = 1$, we obtain

$$y_n = (1 - h\lambda)^{-n}. \tag{4.10}$$

For any stepsize $h > 0$, we have $|1 - h\lambda| > 1$ and so $y_n \to 0$ as $n \to \infty$.

Continuing with Example 4.1, in Table 4.2 we give numerical results for the backward Euler method. A comparison between Tables 4.1 and 4.2 reveals that the backward Euler method is substantially better than the Euler method on the model problem (4.3).

Table 4.2 Backward Euler solution at $x = 0.2$ for Example 4.1

h	$y_h(0.2)$
0.1	8.26e − 3
0.05	7.72e − 4
0.02	1.69e − 5
0.01	9.54e − 7
0.001	5.27e − 9

The major difference between the two methods is that for the backward Euler method, at each timestep, we need to solve a nonlinear algebraic equation

$$y_{n+1} = y_n + h f(t_{n+1}, y_{n+1}) \tag{4.11}$$

for y_{n+1}. Methods in which y_{n+1} must be found by solving a rootfinding problem are called *implicit methods*, since y_{n+1} is defined implicitly. In contrast, methods that give y_{n+1} directly are called *explicit methods*. Euler's method is an explicit method, whereas the backward Euler method is an implicit method. Under the Lipschitz continuity assumption (2.19) on the function $f(t, z)$, it can be shown that if h is small enough, the equation (4.11) has a unique solution.

Traditional rootfinding methods (e.g., Newton's method, the secant method, the bisection method) can be applied to (4.11) to find its root y_{n+1}; but often that is a very time-consuming process. Instead, (4.11) is usually solved by a simple iteration technique. Given an initial guess $y_{n+1}^{(0)} \approx y_{n+1}$, define $y_{n+1}^{(1)}$, $y_{n+1}^{(2)}$, etc., by

$$y_{n+1}^{(j+1)} = y_n + h f(t_{n+1}, y_{n+1}^{(j)}), \qquad j = 0, 1, 2, \ldots . \tag{4.12}$$

It can be shown that if h is sufficiently small, then the iterates $y_{n+1}^{(j)}$ will converge to y_{n+1} as $j \to \infty$. Subtracting (4.12) from (4.11) gives us

$$y_{n+1} - y_{n+1}^{(j+1)} = h \left[f(t_{n+1}, y_{n+1}) - f(t_{n+1}, y_{n+1}^{(j)}) \right],$$

$$y_{n+1} - y_{n+1}^{(j+1)} \approx h \cdot \frac{\partial f(t_{n+1}, y_{n+1})}{\partial y} [y_{n+1} - y_{n+1}^{(j)}].$$

The last formula is obtained by applying the mean value theorem to $f(t_{n+1}, z)$, considered as a function of z. This formula gives a relation between the error in successive iterates. Therefore, if

$$\left| h \cdot \frac{\partial f(t_{n+1}, y_{n+1})}{\partial y} \right| < 1, \tag{4.13}$$

then the errors will converge to zero, as long as the initial guess $y_{n+1}^{(0)}$ is a sufficiently accurate approximation to y_{n+1}.

The preceding iteration method (4.12) and its analysis is a special case of the theory of *fixed-point iteration* for solving a nonlinear equation $z = g(z)$. The iteration scheme is

$$z_{j+1} = g(z_j), \qquad j = 0, 1, 2, \ldots \tag{4.14}$$

with z_0 an initial estimate of the solution being sought. Denote by α the solution we are seeking for the equation $z = g(z)$. Assuming that $g(z)$ is continuously differentiable in a neighborhood of α, we have that the iteration (4.14) will converge if

$$|g'(\alpha)| < 1 \tag{4.15}$$

and if the initial estimate z_0 is chosen sufficiently close to α; see [11, §2.5], [12, §3.4], [68, §6.3]. Applying this notation to our iteration (4.12), $\alpha = y_{n+1}$ is the fixed point, and

$$g(z) \equiv y_n + h f(t_{n+1}, z).$$

The convergence condition (4.13) is simply the condition (4.15).

In practice, one uses a good initial guess $y_{n+1}^{(0)}$, and one chooses an h that is so small that the quantity in (4.13) is much less than 1. Then the error $y_{n+1} - y_{n+1}^{(j)}$ decreases rapidly to a small quantity as j increases, and often only one iterate needs to be computed. The usual choice of the initial guess $y_{n+1}^{(0)}$ for (4.12) is based on the Euler method

$$y_{n+1}^{(0)} = y_n + hf(t_n, y_n). \tag{4.16}$$

This is called a *predictor formula*, as it predicts the root of the implicit method.

For many equations, it is usually sufficient to do the iteration (4.12) once. Thus, a practical way to implement the backward Euler method is to do the following one-point iteration for solving (4.11) approximately:

$$\bar{y}_{n+1} = y_n + h f(t_{n+1}, y_n),$$
$$y_{n+1} = y_n + h f(t_{n+1}, \bar{y}_{n+1}).$$

The resulting numerical method is then given by the formula

$$y_{n+1} = y_n + h f(t_{n+1}, y_n + h f(t_{n+1}, y_n)). \tag{4.17}$$

It can be shown that this method is still of first-order accuracy. However, it is no longer absolutely stable (see Problem 1).

MATLAB® program. We now turn to an implementation of the backward Euler method. At each step, with y_n available from the previous step, we use the Euler method to compute an estimate of y_{n+1}:

$$y_{n+1}^{(1)} = y_n + hf(t_n, y_n).$$

Then we carry out the iteration

$$y_{n+1}^{(k+1)} = y_n + h f(t_{n+1}, y_{n+1}^{(k)})$$

until the difference between successive values of the iterates is sufficiently small, indicating a sufficiently accurate approximation of the solution y_{n+1}. To prevent an infinite loop of iteration, we require the iteration to stop if 10 iteration steps are taken without reaching a satisfactory solution; in this latter case, an error message will be displayed.

```
function [t,y] = euler_back(t0,y0,t_end,h,fcn,tol)
%
% function [t,y] = euler_back(t0,y0,t_end,h,fcn,tol)
%
% Solve the initial value problem
%    y' = f(t,y),  t0 <= t <= b,  y(t0)=y0
% Use the backward Euler method with a stepsize of h.
% The user must supply an m-file to define the
% derivative f, with some name, say 'deriv.m', and a
% first line of the form
%     function ans=deriv(t,y)
% tol is the user supplied bound on the difference
% between successive values of the backward Euler
% iteration.  A sample call would be
%    [t,z]=euler_back(t0,z0,b,delta,'deriv',1.0e-3)
%
% Output:
% The routine euler_back will return two vectors,
% t and y. The vector t will contain the node points
%    t(1)=t0, t(j)=t0+(j-1)*h, j=1,2,...,N
% with
%    t(N) <= t_end,  t(N)+h > t_end
% The vector y will contain the estimates of the
% solution Y at the node points in t.
%

% Initialize.
n = fix((t_end-t0)/h)+1;
t = linspace(t0,t0+(n-1)*h,n)';
y = zeros(n,1);
y(1) = y0;
i = 2;
% advancing
while i <= n
%
% forward Euler estimate
%
   yt1 = y(i-1)+h*feval(fcn,t(i-1),y(i-1));
% one-point iteration
```

```
   count = 0;
   diff = 1;
   while diff > tol & count < 10
      yt2 = y(i-1) + h*feval(fcn,t(i),yt1);
      diff = abs(yt2-yt1);
      yt1 = yt2;
      count = count +1;
   end
   if count >= 10
      disp('Not converging after 10 steps at t = ')
      fprintf('%5.2f\n', t(i))
   end
   y(i) = yt2;
   i = i+1;
end
```

4.2   THE TRAPEZOIDAL METHOD

One main drawback of both the Euler method and the backward Euler method is the low convergence order. Next we present a method that has a higher convergence order and in which, at the same time, the stability property (4.5) is valid for any stepsize h in solving the model problem (4.3).

We begin by introducing the *trapezoidal rule* for numerical integration:

$$\int_a^b g(s)\, ds \approx \tfrac{1}{2}\,(b-a)\,[g(a)+g(b)]. \tag{4.18}$$

This rule is illustrated in Figure 4.1. The graph of $y = g(t)$ is approximated on $[a, b]$ by the linear function $y = p_1(t)$ that interpolates $g(t)$ at the endpoints of $[a, b]$. The integral of $g(t)$ over $[a, b]$ is then approximated by the integral of $p_1(t)$ over $[a, b]$. By using various approaches, we can obtain the more complete result

$$\int_a^b g(s)\, ds = \tfrac{1}{2}\,(b-a)\,[g(a)+g(b)] - \tfrac{1}{12}\,(b-a)^3\, g''(\xi) \tag{4.19}$$

for some $a \le \xi \le b$.

We integrate the differential equation

$$Y'(t) = f(t, Y(t))$$

from t_n to t_{n+1}:

$$Y(t_{n+1}) = Y(t_n) + \int_{t_n}^{t_{n+1}} f(s, Y(s))\, ds. \tag{4.20}$$

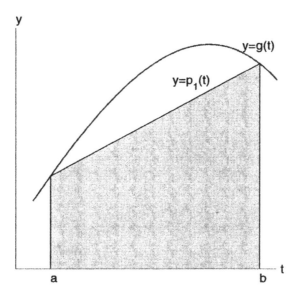

Figure 4.1 Illustration of trapezoidal rule

Use the trapezoidal rule (4.18) to approximate the integral. Applying (4.19) to this integral, we obtain

$$Y(t_{n+1}) = Y(t_n) + \tfrac{1}{2}h\left[f(t_n, Y(t_n)) + f(t_{n+1}, Y(t_{n+1}))\right]$$
$$-\tfrac{1}{12}h^3 Y^{(3)}(\xi_n) \qquad (4.21)$$

for some $t_n \leq \xi_n \leq t_{n+1}$. By dropping the final error term and then equating both sides, we obtain the *trapezoidal method* for solving the initial value problem (1.7):

$$y_{n+1} = y_n + \tfrac{1}{2}h\left[f(t_n, y_n) + f(t_{n+1}, y_{n+1})\right], \quad n \geq 0, \qquad (4.22)$$

with $y_0 = Y_0$.

The truncation error for the trapezoidal method is

$$T_{n+1} = -\tfrac{1}{12}h^3 Y^{(3)}(\xi_n). \qquad (4.23)$$

It can be shown that the trapezoidal method is of second-order accuracy. Assuming $y_0 = Y_0$, we can show

$$\max_{t_0 \leq t_n \leq b} |Y(t_n) - y_h(t_n)| \leq ch^2$$

for all sufficiently small h, with c independent of h. The method of proof is a variation of that used for Euler's method in Chapter 2. In addition, the trapezoidal method is absolutely stable. This higher order and its absolute stability has made the trapezoidal method an important tool when solving partial differential equations of parabolic type; see Section 8.1 in Chapter 8.

Notice that the trapezoidal method is an *implicit method*. In a general step, y_{n+1} is found from the equation

$$y_{n+1} = y_n + \frac{h}{2}[f(t_n, y_n) + f(t_{n+1}, y_{n+1})], \tag{4.24}$$

although this equation can be solved explicitly in only a relatively small number of cases. The discussion for the solution of the backward Euler equation (4.11) applies to the solution of the equation (4.24), with a slight variation. The iteration formula (4.12) is now replaced by

$$y_{n+1}^{(j+1)} = y_n + \frac{h}{2}[f(t_n, y_n) + f(t_{n+1}, y_{n+1}^{(j)})], \qquad j = 0, 1, 2, \dots. \tag{4.25}$$

If $y_{n+1}^{(0)}$ is a sufficiently good estimate of y_{n+1} and if h is sufficiently small, then the iterates $y_{n+1}^{(j)}$ will converge to y_{n+1} as $j \to \infty$. The convergence condition (4.13) is replaced by

$$\left| \frac{h}{2} \cdot \frac{\partial f(t_{n+1}, y_{n+1})}{\partial y} \right| < 1. \tag{4.26}$$

Note that the condition (4.26) is somewhat easier to satisfy than (4.13), indicating that the trapezoidal method is slightly easier to use than the backward Euler method.

The usual choice of the initial guess $y_{n+1}^{(0)}$ for (4.25) is based on the Euler method

$$y_{n+1}^{(0)} = y_n + hf(t_n, y_n), \tag{4.27}$$

or an Adams–Bashforth method of order 2 (see Chapter 6)

$$y_{n+1}^{(0)} = y_n + \frac{h}{2}[3f(t_n, y_n) - f(t_{n-1}, y_{n-1})]. \tag{4.28}$$

These are called *predictor formulas*. In either of these two cases for generating $y_{n+1}^{(0)}$, compute $y_{n+1}^{(1)}$ from (4.25) and accept it as the root y_{n+1}. In the first step ($n = 0$), we use the Euler predictor formula rather than the predictor (4.28). With both methods of choosing $y_{n+1}^{(0)}$, it can be shown that the global error in the resulting solution $\{y_h(t_n)\}$ is still $\mathcal{O}(h^2)$. If the Euler predictor (4.27) is used to define $y_{n+1}^{(0)}$, and if we accept $y_{n+1}^{(1)}$ as the value of y_{n+1}, then the resulting new scheme is

$$y_{n+1} = y_n + \frac{h}{2}[f(t_n, y_n) + f(t_{n+1}, y_n + h f(t_n, y_n))], \tag{4.29}$$

known as *Heun's method*.   The Heun method is still of second-order accuracy. However, it is no longer absolutely stable.

MATLAB program. In our implementation of the trapezoidal method, at each step, with y_n available from the previous step, we use the Euler method to compute an estimate of y_{n+1}:

$$y_{n+1}^{(0)} = y_n + hf(t_n, y_n).$$

Then we use the trapezoidal formula to do the iteration

$$y_{n+1}^{(k+1)} = y_n + \frac{h}{2}\left[f(t_n, y_n) + f(t_{n+1}, y_{n+1}^{(k)})\right]$$

until the difference between successive values of the iterates is sufficiently small, indicating a sufficiently accurate approximation of the solution y_{n+1}. To prevent an infinite loop of iteration, we require the iteration to stop if 10 iteration steps are taken without reaching a satisfactory solution; and in this latter case, an error message will be displayed.

```
function [t,y] = trapezoidal(t0,y0,t_end,h,fcn,tol)
%
% function [t,y] = trapezoidal(t0,y0,t_end,h,fcn,tol)
%
% Solve the initial value problem
%    y' = f(t,y),  t0 <= t <= b,  y(t0)=y0
% Use trapezoidal method with a stepsize of h.  The
% user must supply an m-file to define the derivative
% f, with some name, say 'deriv.m', and a first line
% of the form
%    function ans=deriv(t,y)
% tol is the user supplied bound on the difference
% between successive values of the trapezoidal
% iteration. A sample call would be
%    [t,z]=trapezoidal(t0,z0,b,delta,'deriv',1e-3)
%
% Output:
% The routine trapezoidal will return two vectors,
% t and y. The vector t will contain the node points
%    t(1) = t0, t(j) = t0+(j-1)*h, j=1,2,...,N
% with
%    t(N) <= t_end,  t(N)+h > t_end
% The vector y will contain the estimates of the
% solution Y at the node points in t.
%

% Initialize.
n = fix((t_end-t0)/h)+1;
t = linspace(t0,t0+(n-1)*h,n)';
y = zeros(n,1);
y(1) = y0;
```

```
i = 2;
% advancing
while i <= n
  fyt = feval(fcn,t(i-1),y(i-1));
%
% Euler estimate
%
  yt1 = y(i-1)+h*fyt;
% trapezoidal iteration
  count = 0;
  diff = 1;
  while diff > tol & count < 10
    yt2 = y(i-1) + h*(fyt+feval(fcn,t(i),yt1))/2;
    diff = abs(yt2-yt1);
    yt1 = yt2;
    count = count +1;
  end
  if count >= 10
    disp('Not converging after 10 steps at t = ')
    fprintf('%5.2f\n', t(i))
  end
  y(i) = yt2;
  i = i+1;
end
```

Example 4.2 Consider the problem

$$Y'(t) = \lambda Y(t) + (1 - \lambda) \cos(t) - (1 + \lambda) \sin(t), \qquad Y(0) = 1, \qquad (4.30)$$

whose true solution is $Y(t) = \sin(t) + \cos(t)$. Euler's method is used for the numerical solution, and the results for several values of λ and h are given in Table 4.3. Note that according to the formula (2.10) for the truncation error, we obtain

$$T_{n+1} = \tfrac{1}{2}h^2 Y''(\xi_n).$$

The solution $Y(t)$ does not depend on λ. But the actual global error depends strongly on λ, as illustrated in the table; and the behavior of the global error is directly linked to the size of λh and, thus, to the size of the stability region for Euler's method. The error is small, provided that $|\lambda| h$ is sufficiently small. The cases of an unstable and rapid growth in the error are exactly the cases in which $|\lambda| h$ is outside the range (4.7). We then apply the backward Euler method and the trapezoidal method to the solution of the problem (4.30). The results are shown in Tables 4.4 and 4.5, with the stepsize $h = 0.5$. The error varies with λ, but there are no stability problems, in contrast to the Euler method. The solutions of the backward Euler method and the trapezoidal method for y_{n+1} were done exactly. This is possible because the differential equation is linear in Y. The fixed-point iterations (4.12) and (4.25) do not converge when $|\lambda| h$ is large. ∎

Table 4.3 Euler's method for (4.30)

λ	t	Error $h = 0.5$	Error $h = 0.1$	Error $h = 0.01$
-1	1	$-2.46e - 1$	$-4.32e - 2$	$-4.22e - 3$
	2	$-2.55e - 1$	$-4.64e - 2$	$-4.55e - 3$
	3	$-2.66e - 2$	$-6.78e - 3$	$-7.22e - 4$
	4	$2.27e - 1$	$3.91e - 2$	$3.78e - 3$
	5	$2.72e - 1$	$4.91e - 2$	$4.81e - 3$
-10	1	$3.98e - 1$	$-6.99e - 3$	$-6.99e - 4$
	2	$6.90e + 0$	$-2.90e - 3$	$-3.08e - 4$
	3	$1.11e + 2$	$3.86e - 3$	$3.64e - 4$
	4	$1.77e + 3$	$7.07e - 3$	$7.04e - 4$
	5	$2.83e + 4$	$3.78e - 3$	$3.97e - 4$
-50	1	$3.26e + 0$	$1.06e + 3$	$-1.39e - 4$
	2	$1.88e + 3$	$1.11e + 9$	$-5.16e - 5$
	3	$1.08e + 6$	$1.17e + 15$	$8.25e - 5$
	4	$6.24e + 8$	$1.23e + 21$	$1.41e - 4$
	5	$3.59e + 11$	$1.28e + 27$	$7.00e - 5$

Table 4.4 Backward Euler solution for (4.30); $h = 0.5$

t	Error $\lambda = -1$	Error $\lambda = -10$	Error $\lambda = -50$
2	$2.08e - 1$	$1.97e - 2$	$3.60e - 3$
4	$-1.63e - 1$	$-3.35e - 2$	$-6.94e - 3$
6	$-7.04e - 2$	$8.19e - 3$	$2.18e - 3$
8	$2.22e - 1$	$2.67e - 2$	$5.13e - 3$
10	$-1.14e - 1$	$-3.04e - 2$	$-6.45e - 3$

Equations with λ negative but large in magnitude are examples of *stiff differential equations*. Their truncation error may be satisfactorily small with not too small a value of h, but the large size of $|\lambda|$ may force h to be much smaller in order that λh is in the stability region. The backward Euler method and the trapezoidal method are therefore very desirable because their stability regions contain all λh where λ is negative or λ is complex with negative real part. For stiff differential equations, one must use a numerical method with a large region of absolute stability, or else h must be chosen very small. The backward Euler method is preferred to the trapezoidal method when solving very stiff differential equations (see Problems 14, 15), although

Table 4.5   Trapezoidal solution for (4.30); $h = 0.5$

t	Error $\lambda = -1$	Error $\lambda = -10$	Error $\lambda = -50$
2	$-1.13e - 2$	$-2.78e - 3$	$-7.91e - 4$
4	$-1.43e - 2$	$-8.91e - 5$	$-8.91e - 5$
6	$2.02e - 2$	$2.77e - 3$	$4.72e - 4$
8	$-2.86e - 3$	$-2.22e - 3$	$-5.11e - 4$
10	$-1.79e - 2$	$-9.23e - 4$	$-1.56e - 4$

it is of lower-order. There are other methods, of higher-order, for approximating stiff differential equations (see [44], [72, Chap. 8]); this is an active area of research. More extensive discussions on numerically solving stiff differential equations can be found later in Chapters 8 and 9.

PROBLEMS

1. Show that the method defined by formula (4.17) is not absolutely stable.

2. Show that the trapezoidal method (4.22) is absolutely stable, but the scheme (4.29) is not.

3. Use backward Euler's method to solve Problem 3 of Chapter 2.

4. Use the trapezoidal method to solve Problem 3 of Chapter 2.

5. Apply the backward Euler method to solve the initial value problem in Problem 11 of Chapter 2 for $\alpha = 2.5, 1.5, 1.1$, with $h = 0.2, 0.1, 0.05$. Compute the error in the solution at the nodes, determine the convergence orders numerically, and compare the results with those obtained by Euler's method.

6. Apply the trapezoidal method to solve the initial value problem in Problem 11 of Chapter 2 for $\alpha = 2.5, 1.5, 1.1$, with $h = 0.2, 0.1, 0.05$. Compute the error in the solution at the nodes, determine numerically the convergence orders, and compare the results with that of the Euler method and the backward Euler method.

7. Solve the equation

$$Y'(t) = \lambda Y(t) + \frac{1}{1 + t^2} - \lambda \tan^{-1}(t), \qquad Y(0) = 0;$$

$Y(t) = \tan^{-1}(t)$ is the true solution. Use Euler's method, the backward Euler method, and the trapezoidal method. Let $\lambda = -1, -10, -50$, and $h = 0.5, 0.1, 0.001$. Discuss the results. In implementing the backward Euler method and the trapezoidal method, note that the implicit equation for y_{n+1} can be solved explicitly without iteration.

8. Apply the backward Euler method to the numerical solution of $Y'(t) = \lambda Y(t) + g(t)$ with $\lambda < 0$ and large in magnitude. Investigate how small h must be chosen for the iteration

$$y_{n+1}^{(j+1)} = y_n + h f\left(t_{n+1}, y_{n+1}^{(j)}\right), \qquad j = 0, 1, 2, \dots$$

to converge to y_{n+1}. Is this iteration practical for very large values of $|\lambda|$?

9. Repeat Problem 5 of Chapter 3 using the backward Euler method.

10. Determine whether the midpoint method

$$y_{n+1} = y_n + h f\left(t_{n+1/2}, \tfrac{1}{2}(y_n + y_{n+1})\right),$$

where $t_{n+1/2} = (t_n + t_{n+1})/2$, is absolutely stable.

11. Let $\theta \in [0, 1]$ be a constant, and denote $t_{n+\theta} = (1 - \theta) t_n + \theta t_{n+1}$. Consider the generalized midpoint method

$$y_{n+1} = y_n + h f(t_{n+\theta}, (1 - \theta) y_n + \theta y_{n+1})$$

and its trapezoidal analog

$$y_{n+1} = y_n + h \left[(1 - \theta) f(t_n, y_n) + \theta f(t_{n+1}, y_{n+1})\right].$$

Show that the methods are absolutely stable when $\theta \in [1/2, 1]$. Determine the regions of absolute stability of the methods when $0 \le \theta < \tfrac{1}{2}$.

12. As a special case in which the error of the backward Euler method can be analyzed directly, we consider the model problem (4.3) again, with λ an arbitrary real constant. The backward Euler solution of the problem is given by the formula (4.10). Following the procedure for solving Problem 4 (c) in Chapter 2, show that

$$Y(t_n) - y_h(t_n) = -\frac{\lambda^2 t_n e^{\lambda t_n}}{2} h + \mathcal{O}(h^2).$$

13. Let $Y(t)$ be the solution, if it exists, to the initial value problem (1.7). By integrating, show that Y satisfies

$$Y(t) = Y_0 + \int_{t_0}^{t} f(s, Y(s)) \, ds.$$

Conversely, show that if this equation has a continuous solution on the interval $t_0 \le t \le b$, then the initial value problem (1.7) has the same solution.

14. As in the previous problems, consider the model problem (4.3) with a real constant $\lambda < 0$. Show that the solution of the trapezoidal method is

$$y_h(t_n) = \left(\frac{1 + \frac{1}{2}\lambda h}{1 - \frac{1}{2}\lambda h}\right)^n, \qquad n \geq 0.$$

Rewrite the solution formula as

$$y_h(t_n) = \exp\left(\frac{[\log(1 + \frac{1}{2}\lambda h) - \log(1 - \frac{1}{2}\lambda h)]}{h} t_n\right),$$

and use Taylor polynomial expansions of $\log(1 \pm u)$ about $u = 0$ to show that

$$Y(t_n) - y_h(t_n) = -\frac{1}{12}h^2\lambda^3 t_n e^{\lambda t_n} + \mathcal{O}(h^4).$$

So for h small, the error is almost proportional to h^2.

15. Use the formula (4.10) for the backward Euler method and the formula from Problem 14 for the trapezoidal method to show that the backward Euler method performs better than the trapezoidal method problem (4.3) with λ negatively very large.

16. In this exercise, we consider a method with third-order truncation errors, which is not convergent or stable.

    (a) Given $Y(t)$ 3 times continuously differentiable, show that

    $$Y(t_{n+1}) = 3Y(t_n) - 2Y(t_{n-1}) + \frac{1}{2}h[Y'(t_n) - 3Y'(t_{n-1})]$$
    $$+ \frac{7}{12}h^3 Y'''(t_n) + \mathcal{O}(h^4). \tag{4.31}$$

    Thus a numerical method for solving the differential equation

    $$Y'(t) = f(t, Y(t))$$

    is

    $$y_{n+1} = 3y_n - 2y_{n-1} + \frac{1}{2}h[f(t_n, y_n) - 3f(t_{n-1}, y_{n-1})], \qquad n \geq 1.$$

    This is a numerical method whose truncation error is $\mathcal{O}(h^3)$. It is an example of a multistep method (see Chapter 6). To use the method, we need a value for y_1, called an artificial initial value, in addition to the initial value $y_0 = Y_0$.

    Hint: To prove (4.31), use a quadratic Taylor expansion about the point t_n for $Y(t)$, including an error term $R_3(t)$. Use this to evaluate $Y(t_{n-1})$ and $Y(t_{n+1})$, along with $Y'(t_{n-1})$. Substitute into

    $$Y(t_{n+1}) - \left\{3Y(t_n) - 2Y(t_{n-1}) + \frac{1}{2}h[Y'(t_n) - 3Y'(t_{n-1})]\right\}$$

to obtain the final term in (4.31).

(b) Now apply the method to solve the very simple initial value problem

$$Y'(t) \equiv 0, \qquad Y(0) = 1,$$

whose solution is $Y(t) \equiv 1$. Show that if the initial values are chosen to be $y_0 = 1$, $y_1 = 1 + h$, then the numerical solution is $y_n = 1 - h + h\, 2^n$. Note that $|y_1 - Y(h)| = h \to 0$ as $h \to 0$. Let $t_n = 1$. Show that $|Y(1) - y_n| \to \infty$ as $h \to 0$. Thus, the method is not convergent.

(c) A slight variant of the arguments of (b) can be used to show the instability of the method. Show that with the initial values $y_0 = y_1 = 1$, the numerical solution is $y_n = 1$ for all n, while if the initial values are perturbed to $y_{\epsilon,0} = 1$, $y_{\epsilon,1} = 1 + \epsilon$, then the numerical solution becomes $y_{\epsilon,n} = 1 - \epsilon + \epsilon\, 2^n$. Show that at any fixed node point $t_n = \bar{t} > 0$, $|y_{\epsilon,n} - y_n| \to \infty$ as $h \to 0$. Hence, the method is unstable.

CHAPTER 5

TAYLOR AND RUNGE–KUTTA METHODS

To improve on the speed of convergence of Euler's method, we look for approxima-tions to $Y(t_{n+1})$ that are more accurate than the approximation

$$Y(t_{n+1}) \approx Y(t_n) + hY'(t_n),$$

which led to Euler's method. Since this is a linear Taylor polynomial approximation, it is natural to consider higher-order Taylor approximations. Doing this will lead to a family of methods, called the Taylor methods, depending on the order of the Taylor approximation being used.

In deriving a Taylor method, we need higher-order derivatives of the true solution, and we obtain them using the solution itself by differentiating the differential equation. Such expressions for higher-order derivatives are usually time-consuming. The idea of Runge–Kutta methods is to use combinations of compositions of the right-side function of the equation to approximate the derivative terms to a required order. The resulting Runge–Kutta methods are among the most popular methods in solving initial value problems.

5.1 TAYLOR METHODS

To keep the initial explanations as intuitive as possible, we will develop a Taylor method for the problem

$$Y'(t) = -Y(t) + 2\cos(t), \qquad Y(0) = 1, \tag{5.1}$$

whose true solution is $Y(t) = \sin(t) + \cos(t)$. To approximate $Y(t_{n+1})$ by using information about Y at t_n, use the quadratic Taylor approximation

$$Y(t_{n+1}) \approx Y(t_n) + hY'(t_n) + \tfrac{1}{2}h^2 Y''(t_n). \tag{5.2}$$

Its truncation error is

$$T_{n+1}(Y) = \tfrac{1}{6}h^3 Y'''(\xi_n), \qquad \text{some } t_n \le \xi_n \le t_{n+1}. \tag{5.3}$$

To evaluate the right side of (5.2), we can obtain $Y'(t_n)$ directly from (5.1). For $Y''(t)$, differentiate (5.1) to get

$$Y''(t) = -Y'(t) - 2\sin(t) = Y(t) - 2\cos(t) - 2\sin(t).$$

Then (5.2) becomes

$$Y(t_{n+1}) \approx Y(t_n) + h[-Y(t_n) + 2\cos(t_n)] \\ + \tfrac{1}{2}h^2[Y(t_n) - 2\cos(t_n) - 2\sin(t_n)].$$

By forcing equality, we are led to the numerical method

$$y_{n+1} = y_n + h[-y_n + 2\cos(t_n)] \\ + \tfrac{1}{2}h^2[y_n - 2\cos(t_n) - 2\sin(t_n)], \qquad n \ge 0 \tag{5.4}$$

with $y_0 = 1$. This should approximate the solution of the problem (5.1). Because the truncation error (5.3) contains a higher power of h than was true for Euler's method [see (2.10)], it is hoped that the method (5.4) will converge more rapidly.

Table 5.1 contains numerical results for (5.4) and for Euler's method, and it is clear that (5.4) is superior. In addition, if the results for stepsizes $h = 0.1$ and 0.05 are compared, it can be seen that the errors decrease by a factor of approximately 4 when h is halved. This can be justified theoretically, as is discussed later.

In general, to solve the initial value problem

$$Y'(t) = f(t, Y(t)), \qquad t_0 \le t \le b, \qquad Y(t_0) = Y_0 \tag{5.5}$$

by the Taylor method, select a Taylor approximation of certain order and proceed as described above. For order p, write

$$Y(t_{n+1}) \approx Y(t_n) + hY'(t_n) + \cdots + \frac{h^p}{p!}Y^{(p)}(t_n), \tag{5.6}$$

Table 5.1 Example of second-order Taylor method (5.4)

h	t	$y_h(t)$	Error	Euler Error
0.1	2.0	0.492225829	9.25e − 4	−4.64e − 2
	4.0	−1.411659477	1.21e − 3	3.91e − 2
	6.0	0.682420081	−1.67e − 3	1.39e − 2
	8.0	0.843648978	2.09e − 4	−5.07e − 2
	10.0	−1.384588757	1.50e − 3	2.83e − 2
0.05	2.0	0.492919943	2.31e − 4	−2.30e − 2
	4.0	−1.410737402	2.91e − 4	1.92e − 2
	6.0	0.681162413	−4.08e − 4	6.97e − 3
	8.0	0.843801368	5.68e − 5	−2.50e − 2
	10.0	−1.383454154	3.62e − 4	1.39e − 2

where the truncation error is

$$T_{n+1}(Y) = \frac{h^{p+1}}{(p+1)!} Y^{(p+1)}(\xi_n), \qquad t_n \leq \xi_n \leq t_{n+1}. \tag{5.7}$$

Find $Y''(t), \ldots, Y^{(p)}(t)$ by differentiating the differential equation in (5.5) successively, obtaining formulas that implicitly involve only t_n and $Y(t_n)$. As an illustration, we have the following formulas

$$Y''(t) = f_t + f_y f, \tag{5.8}$$
$$Y^{(3)}(t) = f_{tt} + 2 f_{ty} f + f_{yy} f^2 + f_y(f_t + f_y f), \tag{5.9}$$

where

$$f_t = \frac{\partial f}{\partial t}, \quad f_y = \frac{\partial f}{\partial y}, \quad f_{ty} = \frac{\partial^2 f}{\partial t \partial y},$$

and so on are partial derivatives, and together with f, they are evaluated at $(t, Y(t))$. The formulas for the higher derivatives rapidly become very complicated as the differentiation order is increased.

Substitute these formulas into (5.6) and then obtain a numerical method of the form

$$y_{n+1} = y_n + h y_n' + \frac{h^2}{2} y_n'' + \cdots + \frac{h^p}{p!} y_n^{(p)} \tag{5.10}$$

by forcing (5.6) to be an equality. In the formula,

$$y_n' = f(t_n, y_n), \qquad y_n'' = (f_t + f_y f)(t_n, y_n),$$

and so on, using the pattern of (5.8)–(5.9).

If the solution $Y(t)$ and the derivative function $f(t, z)$ are sufficiently differentiable, then it can be shown that the method (5.10) will satisfy

$$\max_{t_0 \leq t_n \leq b} |Y(t_n) - y_h(t_n)| \leq c h^p \cdot \max_{t_0 \leq t \leq b} \left| Y^{(p+1)}(t) \right|. \tag{5.11}$$

The constant c is similar to that appearing in the error formula (2.20) for Euler's method. A proof can be constructed along the same lines as that used for Theorem 2.4 in Chapter 2. In addition, there is an asymptotic error formula

$$Y(t_n) - y_h(t_n) = h^p D(t_n) + \mathcal{O}(h^{p+1}) \tag{5.12}$$

with $D(t)$ satisfying a certain linear differential equation. The result (5.11) shows that for any integer $p \geq 1$, a numerical method based on the Taylor approximation of order p leads to a convergent numerical method with order of convergence p. The asymptotic result (5.12) justifies the use of Richardson's extrapolation to estimate the error and to accelerate the convergence (see Problems 3, 4).

Example 5.1 With $p = 2$, formula (5.12) leads to

$$Y(t_n) - y_h(t_n) \approx \tfrac{1}{3}[y_h(t_n) - y_{2h}(t_n)]. \tag{5.13}$$

Its derivation is left as Problem 3 for the reader. To illustrate the usefulness of the formula, use the entries from Table 5.1 with $t_n = 10$:

$$y_{0.1}(10) \doteq -1.384588757,$$

$$y_{0.05}(10) \doteq -1.383454154.$$

From (5.13),

$$Y(10) - y_{0.05}(10) \doteq \tfrac{1}{3}[0.001134603] \doteq 3.78 \times 10^{-4}.$$

This is a good estimate of the true error 3.62×10^{-4}, given in Table 5.1. ∎

5.2   RUNGE–KUTTA METHODS

The Taylor method is conceptually easy to work with, but as we have seen, it is tedious and time-consuming to have to calculate the higher-order derivatives. To avoid the need for the higher-order derivatives, the Runge–Kutta methods evaluate $f(t, y)$ at more points, while attempting to retain the accuracy of the Taylor approximation. The methods obtained are fairly easy to program, and they are among the most popular methods for solving the initial value problem.

We begin with Runge–Kutta methods of order 2, and later we consider some higher-order methods. The Runge–Kutta methods have the general form

$$y_{n+1} = y_n + hF(t_n, y_n; h), \qquad n \geq 0, \qquad y_0 = Y_0. \tag{5.14}$$

The quantity $F(t_n, y_n; h)$ can be regarded as some kind of "average slope" of the solution on the interval $[t_n, t_{n+1}]$. But its construction is based on making (5.14) act like a Taylor method. For methods of order 2, we generally choose

$$F(t, y; h) = b_1 f(t, y) + b_2 f(t + \alpha h, \, y + \beta h f(t, y)) \tag{5.15}$$

and determine the constants $\{\alpha, \beta, b_1, b_2\}$ so that when the true solution $Y(t)$ is substituted into (5.14), the truncation error

$$T_{n+1}(Y) \equiv Y(t_{n+1}) - [Y(t_n) + hF(t_n, Y(t_n); h)] \tag{5.16}$$

will satisfy

$$T_{n+1}(Y) = \mathcal{O}(h^3), \tag{5.17}$$

just as with the Taylor method of order 2.

To find the equations for the constants, we use Taylor expansions to compute the truncation error $T_{n+1}(Y)$. For the term $f(t + \alpha h, y + \beta h f(t, y))$, we first expand with respect to the second argument around y. Note that we need a remainder $\mathcal{O}(h^2)$:

$$f(t + \alpha h, y + \beta h f(t, y)) = f(t + \alpha h, y) + f_y(t + \alpha h, y)\beta h f(t, y) + \mathcal{O}(h^2).$$

We then expand the terms with respect to the t variable to obtain

$$f(t + \alpha h, y + \beta h f(t, y)) = f + f_t \alpha h + f_y \beta h f + \mathcal{O}(h^2),$$

where the functions are all evaluated at (t, y). Also, recall from following (5.10) that

$$Y'' = f_t + f_y f.$$

Hence

$$Y(t + h) = Y + hY' + \frac{h^2}{2} Y'' + \mathcal{O}(h^3)$$

$$= Y + hf + \frac{h^2}{2} (f_t + f_y f) + \mathcal{O}(h^3).$$

Then

$$\begin{aligned}
T_{n+1}(Y) &= Y(t + h) - [Y(t) + h\, F(t, Y(t); h)] \\
&= Y + hf + \tfrac{1}{2}h^2(f_t + f_y f) \\
&\quad - [Y + hb_1 f + b_2 h\,(f + \alpha h f_t + \beta h f_y f)] + \mathcal{O}(h^3) \\
&= h\,(1 - b_1 - b_2)\, f + \tfrac{1}{2}h^2[(1 - 2\,b_2\alpha)\, f_t \\
&\qquad\qquad + (1 - 2\,b_2\beta) f_y f] + \mathcal{O}(h^3). \tag{5.18}
\end{aligned}$$

The requirement (5.17) implies that the coefficients must satisfy the system

$$\begin{cases} 1 - b_1 - b_2 = 0, \\ 1 - 2\,b_2\alpha = 0, \\ 1 - 2\,b_2\beta = 0. \end{cases}$$

Therefore

$$b_2 \neq 0, \qquad b_1 = 1 - b_2, \qquad \alpha = \beta = \frac{1}{2b_2}. \tag{5.19}$$

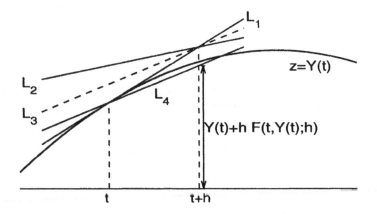

Figure 5.1   An illustration of Runge–Kutta method (5.20); the slope of L_1 is $f(t, Y(t))$, that of L_2 is $f(t + h, Y(t) + hf(t, Y(t)))$, and those of L_3 and L_4 are the average $F(t, Y(t); h)$

Thus there is a family of Runge–Kutta methods of order 2, depending on the choice of b_2. The three favorite choices are $b_2 = \frac{1}{2}, \frac{3}{4}$, and 1.

With $b_2 = \frac{1}{2}$, we obtain the numerical method

$$y_{n+1} = y_n + \frac{h}{2}[f(t_n, y_n) + f(t_n + h, y_n + hf(t_n, y_n))], \qquad n \geq 0. \qquad (5.20)$$

This is also Heun's method (4.29) discussed in Chapter 4. The number $y_n + hf(t_n, y_n)$ is the Euler solution at t_{n+1}. Using it, we obtain an approximation to the derivative at t_{n+1}, namely,

$$f(t_{n+1}, y_n + hf(t_n, y_n)).$$

This and the slope $f(t_n, y_n)$ are then averaged to give an "average" slope of the solution on the interval $[t_n, t_{n+1}]$, giving

$$F(t_n, y_n; h) = \frac{1}{2}[f(t_n, y_n) + f(t_n + h, y_n + hf(t_n, y_n))].$$

This is then used to predict y_{n+1} from y_n, in (5.20). This definition is illustrated in Figure 5.1 for $F(t, Y(t); h)$ as an average slope of Y' on $[t, t + h]$.

Another choice is to use $b_2 = 1$, resulting in the numerical method

$$y_{n+1} = y_n + hf\left(t_n + \tfrac{1}{2}h, y_n + \tfrac{1}{2}hf(t_n, y_n)\right). \qquad (5.21)$$

Table 5.2 Example of second-order Runge–Kutta method

h	t	$y_h(t)$	Error
0.1	2.0	0.491215673	$1.93e-3$
	4.0	-1.407898629	$-2.55e-3$
	6.0	0.680696723	$5.81e-5$
	8.0	0.841376339	$2.48e-3$
	10.0	-1.380966579	$-2.13e-3$
0.05	2.0	0.492682499	$4.68e-4$
	4.0	-1.409821234	$-6.25e-4$
	6.0	0.680734664	$2.01e-5$
	8.0	0.843254396	$6.04e-4$
	10.0	-1.382569379	$-5.23e-4$

Example 5.2 Reconsider the problem (5.1):

$$Y'(t) = -Y(t) + 2\cos(t), \qquad Y(0) = 1.$$

Here

$$f(t, y) = -y + 2\cos(t).$$

The numerical results from using (5.20) are given in Table 5.2. They show that the errors in this Runge–Kutta solution are comparable in accuracy to the results obtained with the Taylor method (5.4). In addition, the errors in Table 5.2 decrease by a factor of approximately 4 when h is halved, confirming the second-order convergence of the method. ∎

5.2.1 A general framework for explicit Runge–Kutta methods

Runge–Kutta methods of higher-order can also be developed. An explicit Runge–Kutta formula with s stages has the following form:

$$
\begin{aligned}
z_1 &= y_n, \\
z_2 &= y_n + ha_{2,1}f(t_n, z_1), \\
z_3 &= y_n + h\left[a_{3,1}f(t_n, z_1) + a_{3,2}f(t_n + c_2 h, z_2)\right], \\
&\vdots \\
z_s &= y_n + h\left[a_{s,1}f(t_n, z_1) + a_{s,2}f(t_n + c_2 h, z_2) \right. \\
&\qquad\qquad \left. + \cdots + a_{s,s-1}f(t_n + c_{s-1} h, z_{s-1})\right],
\end{aligned}
\tag{5.22}
$$

$$
\begin{aligned}
y_{n+1} = y_n + h\left[b_1 f(t_n, z_1) + b_2 f(t_n + c_2 h, z_2) \right. \\
\left. + \cdots + b_{s-1}f(t_n + c_{s-1}h, z_{s-1}) + b_s f(t_n + c_s h, z_s)\right].
\end{aligned}
\tag{5.23}
$$

Here $h = t_{n+1} - t_n$. The coefficients $\{c_i, a_{i,j}, b_j\}$ are given and they define the numerical method. The function F of (5.14), defining a one-step method, is defined implicitly through the formulas (5.22)-(5.23).

More succinctly, we can write the formulas as

$$z_i = y_n + h \sum_{j=1}^{i-1} a_{i,j} f(t_n + c_j h, z_j), \qquad i = 1, \ldots, s, \qquad (5.24)$$

$$y_{n+1} = y_n + h \sum_{j=1}^{s} b_j f(t_n + c_j h, z_j). \qquad (5.25)$$

The coefficients are often displayed in a table called a *Butcher tableau* (after J. C. Butcher):

$$
\begin{array}{c|ccccc}
0 = c_1 & & & & & \\
c_2 & a_{2,1} & & & & \\
c_3 & a_{3,1} & a_{3,2} & & & \\
\vdots & \vdots & & \ddots & & \\
c_s & a_{s,1} & a_{s,2} & \cdots & a_{s,s-1} & \\
\hline
& b_1 & b_2 & \cdots & b_{s-1} & b_s
\end{array}
\qquad (5.26)
$$

The coefficients $\{c_i\}$ and $\{a_{i,j}\}$ are usually assumed to satisfy the conditions

$$\sum_{j=1}^{i-1} a_{i,j} = c_i, \qquad i = 2, \ldots, s. \qquad (5.27)$$

Example 5.3 We give two examples of well-known Runge–Kutta methods.

- The method (5.20) has the Butcher tableau

$$
\begin{array}{c|cc}
0 & & \\
1 & 1 & \\
\hline
& 1/2 & 1/2
\end{array}
$$

- A popular classical method is the following fourth-order procedure.

$$
\begin{aligned}
z_1 &= y_n, \\
z_2 &= y_n + \tfrac{1}{2} h f(t_n, z_1), \\
z_3 &= y_n + \tfrac{1}{2} h f\left(t_n + \tfrac{1}{2}h, z_2\right), \\
z_4 &= y_n + h f\left(t_n + \tfrac{1}{2}h, z_3\right),
\end{aligned}
$$

$$
\begin{aligned}
y_{n+1} = y_n + \tfrac{1}{6} h \big[&f(t_n, z_1) + 2f\left(t_n + \tfrac{1}{2}h, z_2\right) \\
&+ 2f\left(t_n + \tfrac{1}{2}h, z_3\right) + f\left(t_n + h, z_4\right) \big].
\end{aligned}
\qquad (5.28)
$$

The Butcher tableau is

$$
\begin{array}{c|cccc}
0 \\
1/2 & 1/2 \\
1/2 & 0 & 1/2 \\
1 & 0 & 0 & 1 \\
\hline
 & 1/6 & 1/3 & 1/3 & 1/6
\end{array}
\tag{5.29}
$$

Following an extended calculation modeled on that in (5.18), we can show $T_{n+1} = \mathcal{O}(h^5)$.

When the differential equation is simply $Y'(t) = f(t)$ with no dependence of f on Y, this method reduces to Simpson's rule for numerical integration on $[t_n, t_{n+1}]$. The method (5.28) can be easily implemented using a computer or a programmable hand calculator, and it is generally quite accurate. A numerical example is given at the end of the next section. ∎

5.3 CONVERGENCE, STABILITY, AND ASYMPTOTIC ERROR

We want to examine the convergence of the one-step method

$$
y_{n+1} = y_n + hF(t_n, y_n; h), \qquad n \geq 0, \qquad y_0 = Y_0
\tag{5.30}
$$

to the solution $Y(t)$ of the initial value problem

$$
\begin{aligned}
Y'(t) &= f(t, Y(t)), \qquad t_0 \leq t \leq b, \\
Y(t_0) &= Y_0.
\end{aligned}
\tag{5.31}
$$

Using the truncation error of (5.16) for the true solution Y, we introduce

$$
\tau_n(Y) = \frac{1}{h} T_{n+1}(Y).
$$

In order to show convergence of (5.30), we need to have $\tau_n(Y) \to 0$ as $h \to 0$. Since

$$
\tau_n(Y) = \frac{Y(t_{n+1}) - Y(t_n)}{h} - F(t_n, Y(t_n), h; f),
\tag{5.32}
$$

we require that

$$
F(t, Y(t), h; f) \to Y'(t) = f(t, Y(t)) \qquad \text{as } h \to 0.
$$

Accordingly, define

$$
\delta(h) = \sup_{\substack{t_0 \leq t \leq b \\ -\infty < y < \infty}} |f(t, y) - F(t, y, h; f)|,
\tag{5.33}
$$

and assume

$$
\delta(h) \to 0 \qquad \text{as } h \to 0.
\tag{5.34}
$$

This is occasionally called the *consistency condition* for the one-step method (5.30).

We can rewrite (5.32) in the form

$$Y(t_{n+1}) = Y(t_n) + hF(t_n, Y(t_n), h; f) + h\tau_n(Y). \qquad (5.35)$$

We then introduce

$$\tau(h) = \max_{t_0 \le t_n \le b} |\tau_n(Y)|.$$

The condition (5.34) can be used to show $\tau(h) \to 0$ as $h \to 0$; or we may show this result by other means (e.g. see (5.17)).

We also need a Lipschitz condition on F, namely

$$|F(t, y, h; f) - F(t, z, h; f)| \le L|y - z| \qquad (5.36)$$

for all $t_0 \le t \le b$, $-\infty < y, z < \infty$, and all small $h > 0$. This is in analogy with the Lipschitz condition (1.10) for $f(t, z)$ of Chapter 1 which was used to guarantee the existence of a unique solution to the initial value problem for $Y' = f(t, Y)$. The condition (5.36) is usually proved by using the Lipschitz condition (1.10) on $f(t, y)$. For example, with method (5.21), we obtain

$$|F(t, y, h; f) - F(t, z, h; f)|$$
$$= \left| f\left(t + \tfrac{1}{2}h, y + \tfrac{1}{2}hf(t, y)\right) - f\left(t + \tfrac{1}{2}h, z + \tfrac{1}{2}hf(t, z)\right) \right|$$
$$\le K \left| y - z + \tfrac{1}{2}h\left[f(t, y) - f(t, z)\right] \right|$$
$$\le K\left(1 + \tfrac{1}{2}hK\right)|y - z|.$$

The last two inequalities use the Lipschitz condition (1.10) for f. Choose $L = K(1 + \tfrac{1}{2}K)$ for $h \le 1$.

Theorem 5.4 *Assume that the Runge–Kutta method (5.30) satisfies the Lipschitz condition (5.36). Then, for the initial value problem (5.31), the solution $\{y_n\}$ satisfies*

$$\max_{t_0 \le t_n \le b} |Y(t_n) - y_n| \le e^{(b-t_0)L}|Y_0 - y_0| + \left[\frac{e^{(b-t_0)L} - 1}{L}\right]\tau(h), \qquad (5.37)$$

where

$$\tau(h) \equiv \max_{t_0 \le t_n \le b} |\tau_n(Y)|. \qquad (5.38)$$

If the consistency condition (5.34) is also satisfied, then the numerical solution $\{y_n\}$ converges to $Y(t)$.

Proof. Subtract (5.30) from (5.35) to obtain

$$e_{n+1} = e_n + h\left[F(t_n, Y_n, h; f) - F(t_n, y_n, h; f)\right] + h\tau_n(Y) \qquad (5.39)$$

in which $e_n = Y(t_n) - y_n$. Apply the Lipschitz condition (5.36) and use (5.38) to obtain

$$|e_{n+1}| \le (1 + hL)|e_n| + h\tau_n(h), \qquad t_0 \le t_N \le b. \qquad (5.40)$$

As with the convergence proof in Theorem 2.4 for the Euler method, given in Section 2.2 of Chapter 2, this leads easily to the result (5.37).

In most cases, it is known by direct computation that $\tau(h) \to 0$ as $h \to 0$, and in that case, convergence of $\{y_n\}$ to $Y(t)$ is immediately proved. But all that we need to know is that (5.34) is satisfied. To see this, write

$$h\tau_n(Y) = Y(t_{n+1}) - Y(t_n) - hF(t_n, Y(t_n), h; f)$$

$$= hY'(t_n) + \frac{h^2}{2}Y''(\xi_n) - hF(t_n, Y(t_n), h; f),$$

$$h\,|\tau_n(Y)| \le h\delta(h) + \frac{h^2}{2}\,\|Y''\|_\infty,$$

$$\tau(h) \le \delta(h) + \frac{1}{2}h\,\|Y''\|_\infty.$$

Thus $\tau(h) \to 0$ as $h \to 0$, completing the proof. The preceding examples are illustrations of the theorem. ∎

The following result is an immediate consequence of (5.37).

Corollary 5.5 *If the Runge–Kutta method (5.30) has a truncation error $T_n(Y) = \mathcal{O}(h^{m+1})$, then the error in the convergence of $\{y_n\}$ to $Y(t)$ on $[t_0, b]$ is $\mathcal{O}(h^m)$.*

It is not too difficult to derive an asymptotic error formula for the Runge–Kutta method (5.30), provided one is known for the truncation error. Assume

$$T_n(Y) = \varphi(t_n)h^{m+1} + \mathcal{O}(h^{m+2}) \tag{5.41}$$

with $\varphi(t)$ determined by $Y(t)$ and $f(t, Y(t))$. As an example, see the result (5.18) to obtain this expansion for second-order Runge–Kutta methods. Strengthened forms of (5.34) and (5.36) are also necessary. Assume

$$F(t, y, h; f) - F(t, z, h; f) = \frac{\partial F(t, y, h; f)}{\partial y}(y - z) + \mathcal{O}((y - z)^2) \tag{5.42}$$

and also

$$\delta_1(h) \equiv \sup_{\substack{t_0 \le t \le b \\ -\infty < y < \infty}} \left| \frac{\partial f(t, y)}{\partial y} - \frac{\partial F(t, y, h; f)}{\partial y} \right| \to 0 \quad \text{as } h \to 0. \tag{5.43}$$

In practice, both of these results are straightforward to confirm. With these assumptions, we can derive the formula

$$Y(t_n) - y_h(t_n) = D(t_n)h^m + \mathcal{O}(h^{m+1}), \tag{5.44}$$

with $D(t)$ satisfying the linear initial value problem

$$D'(t) = f_y(t, Y(t))D(t) + \varphi(t), \qquad D(t_0) = 0. \tag{5.45}$$

Stability results can be obtained for Runge–Kutta methods in analogy with those for Euler's method as presented in Section 2.4 of Chapter 2. We omit any discussion here.

As with Taylor methods, Richardson's extrapolation can be justified for Runge–Kutta methods using (5.44), and the error can be estimated. For the second-order method (5.20), we obtain the error estimate

$$Y(t_n) - y_h(t_n) \approx \tfrac{1}{3}[y_h(t_n) - y_{2h}(t_n)],$$

just as we obtained it earlier for the second-order Taylor method; see Problem 3.

Example 5.6 Estimate the error for $h = 0.05$ and $t = 10$ in Table 5.2. Then

$$Y(10) - y_{0.05}(10) \doteq \tfrac{1}{3}[-1.3825669379 - (-1.380966579)] \doteq -5.34 \times 10^{-4}.$$

This compares closely with the actual error of -5.23×10^{-1}.   ∎

Example 5.7 Consider the problem

$$Y' = \frac{1}{1 + x^2} - 2Y^2, \qquad Y(0) = 0 \tag{5.46}$$

with the solution $Y = x/(1 + x^2)$. The method (5.28) was used with a fixed stepsize, and the results are shown in Table 5.3. The stepsizes are $h = 0.25$ and $2h = 0.5$. The asymptotic error formula (5.44) becomes

$$Y(x) - y_h(x) = D(x)h^4 + \mathcal{O}(h^5), \tag{5.47}$$

in this case, and this leads to the asymptotic error estimate

$$Y(x) - y_h(x) = \tfrac{1}{15}[y_h(x) - y_{2h}(x)] + \mathcal{O}(h^5). \tag{5.48}$$

In the table the column labeled "Ratio" gives the ratio of the errors for corresponding node points as h is halved. The last column is an example of formula (5.48). Because $T_n(Y) = \mathcal{O}(h^5)$ for method (5.28), Theorem 5.4 implies that the rate of convergence of $y_h(x)$ to $Y(x)$ is $\mathcal{O}(h^4)$. The theoretical value of "Ratio" is 16, and as h decreases further, this value will be realized more closely.   ∎

5.3.1 Error prediction and control

The easiest way to predict the error $Y(t) - y_h(t)$ in a numerical solution $y_h(t)$ is to use Richardson's extrapolation. Solve the initial value problem twice on the given interval $[t_0, b]$, with stepsizes $2h$ and h. Then use Richardson's extrapolation to estimate $Y(t) - y_h(t)$ in terms of $y_h(t) - y_{2h}(t)$, as was done in (5.13) for a second-order method. The cost of estimating the error in this way is an approximately 50% increase in the amount of computation, as compared with the cost of computing just

Table 5.3   Example of Runge-Kutta method (5.28)

x	$y_h(x)$	$Y(x) - y_h(x)$	$Y(x) - y_{2h}(x)$	Ratio	$\frac{1}{15}[y_h(x) - y_{2h}(x)]$
2.0	0.39995699	4.3e − 5	1.0e − 3	24	6.7e − 5
4.0	0.23529159	2.5e − 6	7.0e − 5	28	4.5e − 6
6.0	0.16216179	3.7e − 7	1.2e − 5	32	7.7e − 7
8.0	0.12307683	9.2e − 8	3.4e − 6	36	2.2e − 7
10.0	0.09900987	3.1e − 8	1.3e − 6	41	8.2e − 8

$y_h(t)$. This may seem a large cost, but it is generally worth paying except for the most time-consuming of problems.

It would be desirable to have computer programs that would solve a differential equation on a given interval $[t_0, b]$ with an error less than a given error tolerance $\epsilon > 0$. Unfortunately, this is not possible with most types of numerical methods for the initial value problem. If at some point t we discover that $Y(t) - y_h(t)$ is too large, then the error cannot be reduced by merely decreasing h from that point onward in the computation. The error $Y(t) - y_h(t)$ depends on the cumulative effect of all preceding errors at points $t_n < t$. Thus, to decrease the error at t, it is necessary to repeat the solution of the equation from t_0, but with a smaller stepsize h. For this reason, most package programs for solving the initial value problem will not attempt to directly control the error, although they may try to monitor or bound it. Instead, they use indirect methods to affect the size of the error.

The error $Y(t_n) - y_h(t_n)$ is called the *global error* or total error at t_n. Rather than controlling this global error, we control another error. We introduce the following initial value problem:

$$u'_n(t) = f(t, u_n(t)), \qquad t \geq t_n,$$
$$u_n(t_n) = y_n. \tag{5.49}$$

The solution $u_n(t)$ is called the *local solution* to the differential equation at the point (t_n, y_n). Using it we introduce the *local error*

$$LE_{n+1} = u_n(t_{n+1}) - y_{n+1}. \tag{5.50}$$

This is the error introduced into the solution at the point t_{n+1} when assuming the solution y_n at t_n is the exact solution. Most computer programs that contain error control are based on estimating the local error and then controlling it by varying h suitably. By so doing, they hope to keep the global error sufficiently small. If an error parameter $\epsilon > 0$ is given, the better programs choose the stepsize h to ensure that the local error LE_{n+1} is much smaller, usually satisfying something like

$$|LE_{n+1}| \leq \epsilon(t_{n+1} - t_n). \tag{5.51}$$

This is called controlling the *error per unit stepsize*, with which the global error is generally also kept small. For many differential equations, the global error will then be less than $\epsilon(t_{n+1} - t_0)$.

Table 5.4   Fehlberg coefficients α_i, β_{ij}

i	α_i	β_{i0}	β_{i1}	β_{i2}	β_{i3}	β_{i4}
1	$\frac{1}{4}$	$\frac{1}{4}$				
2	$\frac{3}{8}$	$\frac{3}{32}$	$\frac{9}{32}$			
3	$\frac{12}{13}$	$\frac{1932}{2197}$	$-\frac{7200}{2197}$	$\frac{7296}{2197}$		
4	1	$\frac{439}{216}$	-8	$\frac{3680}{513}$	$-\frac{845}{4104}$	
5	$\frac{1}{2}$	$-\frac{8}{27}$	2	$-\frac{3544}{2565}$	$\frac{1859}{4104}$	$-\frac{11}{40}$

For more detailed discussions of one-step methods, especially Runge–Kutta methods, see Shampine [72], Iserles [48, Chap. 3], and Deuflhard and Bornemann [33, Chaps. 4-6].

5.4   RUNGE–KUTTA–FEHLBERG METHODS

To estimate the local error (5.50), various techniques can be used, including Richardson's extrapolation. A novel technique was devised in the 1970s, and it has led to the currently most popular Runge–Kutta methods. Rather than computing with a method of fixed order, one simultaneously computes by using two methods of different orders. The two methods share most of the function evaluations of f at each step from t_n to t_{n+1}. Then the higher-order formula is used to estimate the error in the lower-order formula. These methods are often called *Fehlberg methods*; we give one such pair of methods, of orders 4 and 5.

Define six intermediate slopes in $[t_n, t_{n+1}]$ by

$$v_0 = f(t_n, y_n),$$

$$v_i = f\left(t_n + \alpha_i h, y_n + h\sum_{j=0}^{i-1}\beta_{ij}v_j\right), \quad i = 1, 2, 3, 4, 5. \tag{5.52}$$

Then the fourth- and fifth-order formulas are given by

$$y_{n+1} = y_n + h\sum_{i=0}^{4}\gamma_i v_i, \tag{5.53}$$

$$\hat{y}_{n+1} = y_n + h\sum_{i=0}^{5}\delta_i v_i. \tag{5.54}$$

The coefficients $\alpha_i, \beta_{ij}, \gamma_i, \delta_i$ are given in Tables 5.4 and 5.5.

The local error in the fourth-order formula (5.53) is estimated by

$$LE_{n+1} \approx \hat{y}_{n+1} - y_{n+1}. \tag{5.55}$$

Table 5.5 Fehlberg coefficients γ_i, δ_i

i	0	1	2	3	4	5
γ_i	$\frac{25}{216}$	0	$\frac{1408}{2565}$	$\frac{2197}{4104}$	$-\frac{1}{5}$	
δ_i	$\frac{16}{135}$	0	$\frac{6656}{12825}$	$\frac{28561}{56430}$	$-\frac{9}{50}$	$\frac{2}{55}$

Table 5.6 Example of fourth-order Fehlberg formula (5.53)

h	t	$y_h(t)$	$Y(t) - y_h(t)$	$\hat{y}_h(t) - y_h(t)$
0.25	2.0	0.493156301	$-5.71e-6$	$-9.49e-7$
	4.0	-1.410449823	$3.71e-6$	$1.62e-6$
	6.0	0.680752304	$2.48e-6$	$-3.97e-7$
	8.0	0.843864007	$-5.79e-6$	$-1.29e-6$
	10.0	-1.383094975	$2.34e-6$	$1.47e-6$
0.125	2.0	0.493150889	$-2.99e-7$	$-2.35e-8$
	4.0	-1.410446334	$2.17e-7$	$4.94e-8$
	6.0	0.680754675	$1.14e-7$	$-1.76e-8$
	8.0	0.843858525	$-3.12e-7$	$-3.47e-8$
	10.0	-1.383092786	$1.46e-7$	$4.65e-8$

It can be shown that this is a correct asymptotic result as $h \to 0$. By using this estimate, if LE_{n+1} is too small or too large, the stepsize can be varied so as to give a value for LE_{n+1} of acceptable size. Note the two formulas (5.53) and (5.54) use the common intermediate slopes $v_0, \ldots, v_4$. At each step, we need to evaluate only six intermediate slopes. In a number of programs, the fifth-order solution $\hat{y}_{n+1}$ is actually the numerical solution used, even though the error is being controlled only for the fourth-order solution y_{n+1}.

Example 5.8 Solve

$$Y'(t) = -Y(t) + 2\cos(t), \qquad Y(0) = 1 \qquad (5.56)$$

whose true solution is $Y(t) = \sin(t) + \cos(t)$. Table 5.6 contains numerical results for $h = 0.25$ and 0.125. Compare the global errors with those in Tables 5.1 and 5.2, where second-order methods are used. Also, it can be seen that the global errors in y_h decrease by factors of 17 to 21, which are fairly close to the theoretical value of 16 for a fourth-order method. The truncation errors, estimated from (5.55), are included to show that they are quite different from the global error. The preceding examples are illustrations of the theorem. ∎

The method (5.52) to (5.55) uses $\hat{y}_{n+1}$ only for estimating the truncation error in the fourth-order method. In practice, $\hat{y}_{n+1}$ is kept as the numerical solution rather than y_{n+1}; thus $\hat{y}_n$ should replace y_n on the right sides of (5.52) to (5.54). The quantity

Table 5.7 Example of fifth-order method (5.54)

h	t	$\hat{y}_n(t)$	$Y(t) - \hat{y}_n(t)$
0.25	2.0	0.493151148	$-5.58e - 7$
	4.0	-1.410446359	$2.43e - 7$
	6.0	0.680754463	$3.26e - 7$
	8.0	0.843858731	$-5.18e - 7$
	10.0	-1.383092745	$1.05e - 7$
0.125	2.0	0.493150606	$-1.61e - 8$
	4.0	-1.410446124	$8.03e - 9$
	6.0	0.680754780	$8.65e - 9$
	8.0	0.843858228	$1.53e - 8$
	10.0	-1.383092644	$4.09e - 9$

in (5.55) will still be the truncation error in the fourth-order method. Programs based on this will be fifth-order, but they will vary their stepsize h to control the local error in the fourth-order method. This tends to make these programs very accurate with regard to global error.

Example 5.9 Repeat the last example, but use the fifth-order method described in the preceding paragraph. The results are given in Table 5.7. Note that the errors decrease by approximately 32 when h is halved, consistent with a fifth-order method. ∎

5.5 MATLAB CODES

MATLAB® contains an excellent suite of programs for solving the initial value problem for systems of ordinary differential equations and related problems. The programs use a variety of methods, and in this text we introduce and illustrate a few of these programs. For a complete description of these programs and the various options that are available when using them, go to the documentation for MATLAB or to the excellent text by Shampine et al. [74]. Each such MATLAB program solves a given differential equation in such a manner that the estimated local error in each component of the solution satisfies a given error test. For a single equation the estimated local error in passing from $y(t_n)$ to $y(t_{n+1})$, call it $e(t_n)$, is to satisfy

$$|e(t_n)| \leq \max \left\{ AbsTol, RelTol \cdot |y(t_n)| \right\}.$$

The error tolerances *AbsTol* and *RelTol* can be specified by having the user run the MATLAB program odeset; when left unspecified, the default tolerances are $AbsTol = 10^{-6}$, $RelTol = 10^{-3}$. For a discussion of the construction of this MAT-LAB suite for solving ordinary differential equations, see Shampine and Reichelt [73] or Shampine, Gladwell, and Thompson [74].

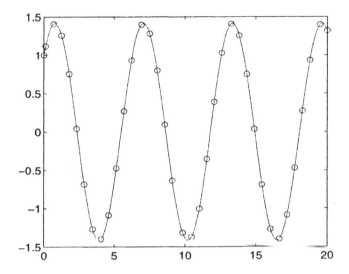

Figure 5.2 The solution values to (5.56) obtained by ode45 are indicated by the symbol o. The curve line is obtained by interpolating these solution values from ode45 using deval

The code ode45 is an implementation of a method similar to the Runge–Kutta–Fehlberg method presented earlier. The program ode45 uses a pair of formulas of orders 4 and 5 by Dormand and Prince [34, cf. Table 2], again estimating the local error as in (5.55). We illustrate the use of ode45 with the following program test_ode45.

Example 5.10 We illustrate the use of ode45 by solving the earlier test equation (5.56). When calling test_ode45, we use $\lambda = -1$ and the error tolerances $AbsTol = 10^{-6}, RelTol = 10^{-4}$. In the program test_ode45, odeset is used to set parameter values that are used in ode45. For a complete description of these parameter values and for more a complete discussion of the varied options for using ode45, consult the MATLAB documentation. We note that in the call to program ode45, we specify the derivative function by giving as an input the function handle @deriv. The output *soln* from ode45 is a MATLAB structure, and it contains all of the information needed to obtain the solution and to interpolate the solution to other values of the independent variable. In our test program, we use the MATLAB program deval to carry out the interpolation on an evenly spaced grid. This could have been done directly when calling ode45, but we have chosen a more general approach to using ode45. Figures 5.2 and 5.3 contain, respectively, the interpolated numerical solution and the error in it. ∎

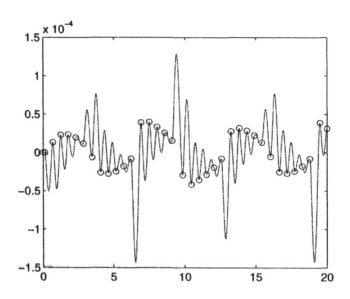

Figure 5.3 The errors in the solution to (5.56) obtained using ode45

The code described in Example 5.10 proceeds as follows.

```
function test_ode45(lambda,relerr,abserr)
%
% function test_ode45(lambda,relerr,abserr)
%
% This is a test program for the ode solver 'ode45'.
% The test is carried out for the single equation
% y' = lambda*y + (1-lambda)*cos(t) - (1+lambda)*sin(t)
% The initial value at t=0 is y(0)=1.  The true solution is
%    y = cos(t) + sin(t)
% The user can input the relative and absolute error
% tolerances to be used by ode45.  These are incorporated
% using the initialization program 'odeset'.
% The program can be adapted easily to other equations and
% other parameter values.

% Initialize and solve
options = odeset('RelTol',relerr,'AbsTol',abserr);
t_begin = 0; t_end = 20;
y_initial = true_soln(t_begin);
num_fcn_eval = 0; % initialize count of derivative evaluations
soln = ode45(@deriv,[t_begin,t_end],y_initial,options);
```

```
                    % See below for function deriv.

% Produce the solution on a uniform grid using interpolation
% of the solution obtained by ode45.  The points plotted with
% 'o' are for the node points returned by ode45.
h_plot = (t_end-t_begin)/200; t_plot = t_begin:h_plot:t_end;
y_plot = deval(soln,t_plot);
figure
plot(soln.x,soln.y,'o',t_plot,y_plot)
title(['Interpolated solution:',...
       ' points noted by ''o'' are at ode45 solution nodes'])
xlabel(['\lambda = ',num2str(lambda)])
disp('press on any key to continue')
pause

% Produce the error in the solution on the uniform grid.
% The points plotted with 'o' are for the solution values
% at the points returned by ode45.
y_true = true_soln(t_plot);
error = y_true - y_plot;
y_true_nodes = true_soln(soln.x);
error_nodes = y_true_nodes - soln.y;
figure
plot(soln.x,error_nodes,'o',t_plot,error)
title('Error in interpolated solution')
xlabel(['\lambda = ',num2str(lambda)])

norm_error = norm(error,inf);
disp(['maximum of error = ',num2str(norm_error)])
disp(['number of derivative evaluations = ',...
      num2str(num_fcn_eval)])

function dy = deriv(t,y)

% Define the derivative in the differential equation.
dy = lambda*y + (1-lambda)*cos(t) - (1+lambda)*sin(t);
num_fcn_eval = num_fcn_eval + 1;
end % deriv

function true = true_soln(t)

% Define the true solution of the initial value problem.
true = sin(t) + cos(t);
end % true_soln
end % test_ode45
```

5.6 IMPLICIT RUNGE–KUTTA METHODS

Return to (5.24)–(5.25) for the definition of an s-stage Runge–Kutta (RK) method. An s-stage *implicit Runge–Kutta method* has the form

$$z_i = y_n + h \sum_{j=1}^{s} a_{i,j} f(t_n + c_j h, z_j), \qquad i = 1, \ldots, s, \qquad (5.57)$$

$$y_{n+1} = y_n + h \sum_{j=1}^{s} b_j f(t_n + c_j h, z_j). \qquad (5.58)$$

It has the Butcher tableau

$$
\begin{array}{c|ccc}
c_1 & a_{1,1} & \cdots & a_{1,s} \\
c_2 & a_{2,1} & \cdots & a_{2,s} \\
\vdots & \vdots & & \vdots \\
c_s & a_{s,1} & \cdots & a_{s,s} \\
\hline
 & b_1 & \cdots & b_s
\end{array}
\qquad (5.59)
$$

We give here a very brief introduction to implicit RK methods, referring to Chapter 9 for a more extensive discussion of the topic.

The equations (5.57) form a simultaneous system of s nonlinear equations for the s unknowns $z_1, \ldots, z_s$; and if the equation $y' = f(t, y)$ is a system of m differential equations, then (5.57) is a simultaneous system of sm nonlinear scalar equations. Why does one want to consider such a complicated numerical method? The answer is that a number of such methods (5.57)-(5.58) have desirable numerical stability properties that are important in solving a variety of important classes of differential equations.

We introduce one approach to deriving many such methods. We begin by converting the differential equation

$$Y'(t) = f(t, Y(t))$$

into an integral equation. Integrating the equation over the interval $[t_n, t]$, we obtain

$$\int_{t_n}^{t} Y'(r)\, dr = \int_{t_n}^{t} f(r, Y(r))\, dr,$$

$$Y(t) = Y(t_n) + \int_{t_n}^{t} f(r, Y(r))\, dr. \qquad (5.60)$$

Approximate the equation, first by replacing $Y(t_n)$ with y_n, and then by replacing the integrand with a polynomial interpolant of it. In particular, choose a set of parameters

$$0 \le \tau_1 < \cdots < \tau_s \le 1.$$

Let $p(r)$ be the unique polynomial of degree $< s$ that interpolates $f(r, Y(r))$ at the node points $\{t_{n,i} \equiv t_n + \tau_i h : i = 1, \ldots, s\}$ on $[t_n, t_{n+1}]$; see Appendix B. Then

(5.60) is approximated by

$$Y(t) \approx y_n + \int_{t_n}^{t} p(r) \, dr. \tag{5.61}$$

Using the Lagrange form of the interpolation polynomial [see (B.6) from Appendix B], we write

$$p(r) = \sum_{j=1}^{s} f(t_{n,j}, Y(t_{n,j})) l_j(r).$$

The Lagrange basis functions $\{l_j(r)\}$ can be obtained from (B.4). Then (5.61) becomes

$$Y(t) \approx y_n + \sum_{j=1}^{s} f(t_{n,j}, Y(t_{n,j})) \int_{t_n}^{t} l_j(r) \, dr. \tag{5.62}$$

We now determine approximate values for $\{Y(t_{n,j}) : j = 1, \ldots, s\}$ by forcing equality in the expression (5.62) at the points $\{t_{n,j}\}$. Let $\{y_{n,j}\}$ denote these approximate values. They are to be determined by solving the nonlinear system

$$y_{n,i} = y_n + \sum_{j=1}^{s} f(t_{n,j}, y_{n,j}) \int_{t_n}^{t_{n,i}} l_j(r) \, dr, \qquad i = 1, \ldots, s. \tag{5.63}$$

If $\tau_s = 1$, then we define $y_{n+1} = y_{n,s}$. Otherwise, we define

$$y_{n+1} = y_n + \sum_{j=1}^{s} f(t_{n,j}, y_{n,j}) \int_{t_n}^{t_{n+1}} l_j(r) \, dr. \tag{5.64}$$

The integrals in (5.63) and (5.64) are easily evaluated, and we will give a particular case below with $s = 2$.

The general method of forcing an approximating equation to be true at a given set of node points is called *collocation*, and the points $\{t_{n,i}\}$ at which equality is forced are called the *collocation node points*. We should note that some Runge–Kutta methods are not collocation methods. An example is the following implicit method given by Iserles [48, p. 44]:

$$\begin{array}{c|cc}
0 & 0 & 0 \\
2/3 & 1/3 & 1/3 \\
\hline
 & 1/4 & 3/4
\end{array} \tag{5.65}$$

5.6.1 Two-point collocation methods

Let $0 \le \tau_1 < \tau_2 \le 1$, and recall that $t_{n,1} = t_n + h\tau_1$ and $t_{n,2} = t_n + h\tau_2$. Then the interpolation polynomial is

$$p(r) = \frac{1}{h(\tau_2 - \tau_1)} [(t_{n+1} - r) f(t_{n,1}, Y(t_{n,1})) + (r - t_n) f(t_{n,2}, Y(t_{n,2}))]. \tag{5.66}$$

Following calculation of the integrals, the system (5.64) has the Butcher tableau

$$
\begin{array}{c|cc}
\tau_1 & (\tau_2^2 - [\tau_2 - \tau_1]^2)/(2[\tau_2 - \tau_1]) & -\tau_1^2/(2[\tau_2 - \tau_1]) \\
\tau_2 & \tau_2^2/(2[\tau_2 - \tau_1]) & ([\tau_2 - \tau_1]^2 - \tau_1^2)/(2[\tau_2 - \tau_1]) \\
\hline
 & (\tau_2^2 - [1 - \tau_2]^2)/(2[\tau_2 - \tau_1]) & ([1 - \tau_1]^2 - \tau_1^2)/(2[\tau_2 - \tau_1])
\end{array}
\tag{5.67}
$$

As a special case, note that when $\tau_1 = 0$ and $\tau_2 = 1$, the system (5.64) becomes

$$
\begin{aligned}
y_{n,1} &= y_n, \\
y_{n,2} &= y_n + \tfrac{1}{2}h\left[f(t_n, y_{n,1}) + f(t_{n+1}, y_{n,2})\right].
\end{aligned}
$$

Substituting from the first equation into the second equation and using $y_{n+1} = y_{n,2}$, we have

$$
y_{n+1} = y_n + \tfrac{1}{2}h\left[f(t_n, y_n) + f(t_{n+1}, y_{n+1})\right],
$$

which is simply the trapezoidal method.

Another choice that has very good convergence and stability properties is to use

$$
\tau_1 = \tfrac{1}{2} - \tfrac{1}{6}\sqrt{3}, \qquad \tau_2 = \tfrac{1}{2} + \tfrac{1}{6}\sqrt{3}.
\tag{5.68}
$$

The Butcher tableau is

$$
\begin{array}{c|cc}
(3 - \sqrt{3})/6 & 1/4 & (3 - 2\sqrt{3})/12 \\
(3 + \sqrt{3})/6 & (3 + 2\sqrt{3})/12 & 1/4 \\
\hline
 & 1/2 & 1/2
\end{array}
\tag{5.69}
$$

The associated nonlinear system is

$$
y_{n,i} = y_n + \sum_{j=1}^{2} a_{i,j} f(t_n + \tau_j h, y_{n,j}), \qquad i = 1, 2,
\tag{5.70}
$$

where we have used the implicit definition of $\{a_{i,j}\}$ that uses (5.59) to reference the elements in (5.69). Then

$$
y_{n+1} = y_n + \frac{h}{2}\left[f(t_{n+1}, y_{n,1}) + f(t_{n+1}, y_{n,2})\right].
\tag{5.71}
$$

This method, called the *two stage Gauss method*, is exact for all polynomial solutions $Y(t)$ of degree ≤ 4. Showing that it has degree of precision 2 is straightforward, because the linear interpolation formula (5.66) is exact when $Y'(t) = f(t, Y(t))$ is linear. Proving that the degree of precision is 4 is a more substantial argument, and we refer the reader to [48, p. 46]. It can be shown that the truncation error for this method has size $\mathcal{O}(h^5)$, and thus the convergence is $\mathcal{O}(h^4)$. It also has desirable stability properties, some of which are taken up in Problem 15 and some of which are deferred to Chapter 9. A disadvantage of the method is the need to solve the nonlinear system in (5.70).

A number of other families of implicit Runge–Kutta methods are discussed in Chapter 9. These methods have stability properties that make them especially useful for solving stiff differential equations.

PROBLEMS

1. A Taylor method of order 3 for problem (5.1) can be obtained using the same procedure that led to (5.4). On the basis of third-order Taylor approximation

$$Y(t_{n+1}) \approx Y(t_n) + hY'(t_n) + \frac{h^2}{2}Y''(t_n) + \frac{h^3}{6}Y'''(t_n),$$

derive the numerical method

$$y_{n+1} = y_n + h[-y_n + 2\cos(t_n)] + \frac{h^2}{2}[y_n - 2\cos(t_n) - 2\sin(t_n)]$$

$$+ \frac{h^3}{6}[-y_n + 2\sin(t_n)], \qquad n \geq 0. \tag{5.72}$$

Implement the numerical method (5.72) for solving the problem (5.1). Compute with stepsizes of $h = 0.1, 0.05$ for $0 \leq t \leq 10$. Compare to the values in Table 5.1, and also check the ratio by which the error decreases when h is halved.

Hint: To simplify the programming, just modify the Euler program given in Chapter 2.

2. Compute solutions to the following problems with a second-order Taylor method. Use stepsizes $h = 0.2, 0.1, 0.05$.

   (a) $Y'(t) = [\cos(Y(t))]^2, \quad 0 \leq t \leq 10, \quad Y(0) = 0;$
   $Y(t) = \tan^{-1}(t).$

   (b) $Y'(t) = 1/(1+t^2) - 2[Y(t)]^2, \quad 0 \leq t \leq 10, \quad Y(0) = 0;$
   $Y(t) = t/(1+t^2).$

   (c) $Y'(t) = \frac{1}{4}Y(t)[1 - \frac{1}{20}Y(t)], \quad 0 \leq t \leq 20, \quad Y(0) = 1;$
   $Y(t) = 20/(1 + 19e^{-t/4}).$

   (d) $Y'(t) = -[Y(t)]^2, \quad 1 \leq t \leq 10, \quad Y(1) = 1;$
   $Y(t) = 1/t.$

   (e) $Y'(t) = -e^{-t}Y(t), \quad 0 \leq t \leq 10, \quad Y(0) = 1;$
   $Y(t) = \exp(e^{-t} - 1).$

These were solved previously in Problems 1 and 2 of Chapter 2. Compare your results with those earlier ones.

3. Recall the asymptotic error for Taylor methods, given in (5.12). For second-order methods, this yields

$$Y(t_n) - y_h(t_n) = h^2 D(t_n) + \mathcal{O}(h^3).$$

From this, derive the Richardson extrapolation formula

$$Y(t_n) = \tfrac{1}{3}[4y_h(t_n) - y_{2h}(t_n)] + \mathcal{O}(h^3)$$

$$\approx \tfrac{1}{3}[4y_h(t_n) - y_{2h}(t_n)] \equiv \widetilde{y}_h(t_n)$$

and the asymptotic error estimate

$$Y(t_n) - y_h(t_n) = \tfrac{1}{3}[y_h(t_n) - y_{2h}(t_n)] + \mathcal{O}(h^3)$$

$$\approx \tfrac{1}{3}[y_h(t_n) - y_{2h}(t_n)].$$

Hint: Consider the formula

$$Y(t_n) - y_{2h}(t_n) = 4h^2 D(t_n) + \mathcal{O}(h^3)$$

and combine it suitably with the earlier formula for $Y(t_n) - y_h(t_n)$.

4. Repeat Problem 3 for methods of a general order $p \geq 1$. Derive the formulas

$$Y(t_n) \approx \frac{1}{2^p - 1}[2^p y_h(t_n) - y_{2h}(t_n)] \equiv \widetilde{y}_h(t_n)$$

with an error proportional to h^{p+1}, and

$$Y(t_n) - y_h(t_n) \approx \frac{1}{2^p - 1}[y_h(t_n) - y_{2h}(t_n)].$$

5. Use Problem 3 to estimate the errors in the results of Table 5.1, for $h = 0.05$. Also produce the Richardson extrapolate $\widetilde{y}_h(t_n)$ and calculate its error. Compare its accuracy to that of $y_h(t_n)$.

6. Derive the second-order Runge–Kutta methods (5.14) corresponding to $b_2 = \tfrac{3}{4}$ and $b_2 = 1$ in (5.15). For $b_2 = 1$, draw an illustrative graph analogous to that of Figure 5.1 for $b_2 = \tfrac{1}{2}$. Give the Butcher tableau for this method.

7. Give the Butcher tableau for each of the following methods.
   (a) The second-order method (5.21)
   (b) The Fehlberg formulas (5.53) and (5.54).

8. Solve the problem (5.1) with one of the formulas from Problem 6. Compare your results to those in Table 5.2 for formula (5.20) with $b_2 = \tfrac{1}{2}$.

9. Using (5.20), solve the equations in Problem 2. Estimate the error by using Problem 3, and compare it to the true error.

10. Implement the classical procedure (5.28), and apply it to the equation (5.1). Solve it with stepsizes of $h = 0.25$ and 0.125. Compare with the results in Table 5.6, the fourth-order Fehlberg example.

    Hint: Modify the Euler program of Chapter 2.

11. Use the program of Problem 10 to solve the equations in Problem 2.

12. Modify the Euler program of Chapter 3 to implement the Runge–Kutta method given in (3.26). With this program, repeat Problems 5 and 6 of Chapter 3.

13. Consider the predator-prey model of (3.4), with the particular constants $A = 4$, $B = 0.5$, $C = 3$, and $D = \frac{1}{3}$. Also, recall Problem 8 in Chapter 3.

    (a) Show that there is a solution $Y_1(t) = C_1$, $Y_2(t) = C_2$, with C_1 and C_2 nonzero constants. What would be the physical interpretation of such a solution $Y(t)$?
    Hint: What are $Y_1'(t)$ and $Y_2'(t)$ in this case?

    (b) Solve this system (3.4) with $Y_1(0) = 3$, $Y_2(0) = 5$, for $0 \le t \le 4$, and use the Runge–Kutta method of Problem 12 with stepsizes of $h = 0.01$ and 0.005. Examine and plot the values of the output in steps of t of 0.1. In addition to these plots of t vs. $Y_1(t)$ and t vs. $Y_2(t)$, also plot Y_1 vs. Y_2.

    (c) Repeat (b) for the initial values $Y_1(0) = 3$, $Y_2(0) = 1, 1.5, 1.9$ in succession. Comment on the relation of these solutions to one another and to the solution of part (a).

14. Show that the implicit Runge–Kutta method (5.65) has a truncation error of size $\mathcal{O}(h^3)$. This can then be used to prove that the method has order of convergence 2.

15. Apply the implicit Runge–Kutta method (5.69) to the model problem

$$Y' = \lambda Y, \qquad t \ge 0,$$
$$Y(0) = 1.$$

    (a) Show that the solution can be written as $y_n = [R(\lambda h)]^n$ with

$$R(z) = \frac{1 + \frac{1}{2}z + \frac{1}{12}z^2}{1 - \frac{1}{2}z + \frac{1}{12}z^2}.$$

    (b) For any real $z < 0$ show that $|R(z)| < 1$. In fact, this bound is true for any complex z with $\mathrm{Real}(z) < 0$, and this implies that the method is absolutely stable.

16. Solve the equations of Problem 2 with the built-in ode45 function. Experiment with several choices of error tolerances, including an absolute error tolerance of $AbsTol = 10^{-4}$ and $\epsilon = 10^{-6}$, along with a relative error tolerance of $RelTol = 10^{-8}$.

17. Solve the equations of Problem 2 with the built-in ode23 function. Experiment with several choices of error tolerances, including an absolute error tolerance of $AbsTol = 10^{-4}$ and $\epsilon = 10^{-6}$, along with a relative error tolerance of $RelTol = 10^{-8}$.

18. Repeat Problem 13 using ode45.

19. Consider the motion of a particle of mass m falling vertically under the earth's gravitational field, and suppose that the downward motion is opposed by a frictional force $p(v)$ dependent on the velocity $v(t)$ of the particle. Then the velocity satisfies the equation

$$mv'(t) = -mg + p(v), \qquad t \geq 0, \qquad v(0) \text{ given}.$$

Let $m = 1$ kg, $g = 9.8$ m/s^2, and $v(0) = 0$. Solve the differential equation for $0 \leq t \leq 20$ and for the following choices of $p(v)$:

(a) $p(v) = -0.1v$, which is positive for a falling body.

(b) $p(v) = 0.1v^2$.

Find answers to at least three digits of accuracy. Graph the functions $v(t)$. Compare the solutions.

20. Consider solving the initial value problem

$$Y'(t) = t - Y(t)^2, \qquad Y(0) = 0$$

on the interval $0 \leq t \leq 20$. Create a Taylor series method of order 2. Implement it in MATLAB and use stepsizes of $h = 0.4, 0.2$, and 0.1 to solve for an approximation to Y. Estimate the error by using Problem 3. Graph the solution that you obtain.

21. Repeat Problem 20 with various initial values $Y(0)$. In particular, use $Y(0) = -0.2, -0.4, -0.6, -0.8$. Comment on your results.

22. Repeat Problems 20 and 21, but use a second-order Runge–Kutta method.

23. Repeat Problems 20 and 21, but use the MATLAB code ode45. Do not attempt to estimate the error since that is embedded in ode113.

24. Consider the problem

$$Y' = \frac{1}{t+1} + c \cdot \tan^{-1}(Y(t)) - \frac{1}{2}, \qquad Y(0) = 0$$

with c a given constant. Since $Y'(0) = \frac{1}{2}$, the solution $Y(t)$ is initially increasing as t increases, regardless of the value of c. As best you can, show that there is a value of c, call it c^*, for which (1) if $c > c^*$, the solution $Y(t)$ increases indefinitely, and (2) if $c < c^*$, then $Y(t)$ increases initially, but then peaks and decreases. Using `ode45`, determine c^* to within 0.00005, and then calculate the associated solution $Y(t)$ for $0 \leq t \leq 50$.

25. (a) Using the Runge–Kutta method (5.20), solve

$$Y'(t) = -Y(t) + t^{0.1}(1.1 + t), \qquad Y(0) = 0,$$

whose solution is $Y(t) = t^{1.1}$. Solve the equation on $[0, 5]$, printing the solution and the errors at $t = 1, 2, 3, 4, 5$. Use stepsizes $h = 0.1, 0.05, 0.025, 0.0125, 0.00625$. Calculate the ratios by which the errors decrease when h is halved. How does this compare with the theoretical rate of convergence of $\mathcal{O}(h^2)$. Explain your results as best you can.

(b) What difficulty arises in attempting to use a Taylor method of order ≥ 2 to solve the equation of part (a)? What does it tell us about the solution?

26. Consider the three-stage Runge–Kutta formula

$$z_1 = y_n,$$
$$z_2 = y_n + ha_{2,1}f(t_n, z_1),$$
$$z_3 = y_n + h\left[a_{3,1}f(t_n, z_1) + a_{3,2}f(t_n + c_2h, z_2)\right],$$
$$y_{n+1} = y_n + h\left[b_1 f(t_n, z_1) + b_2 f(t_n + c_2h, z_2) + b_3 f(t_n + c_3h, z_3)\right].$$

Generalize the argument used in (5.14)–(5.19) for determining the two-stage Runge–Kutta formulas of order 2. Determine the set of equations that the coefficients $\{b_j, c_j, a_{ij}\}$ must satisfy if the formula given above is to be of order 3. Find a particular solution to these equations.

CHAPTER 6

MULTISTEP METHODS

Taylor methods and Runge–Kutta (RK) methods are known as *single-step* or *one-step methods*, since at a typical step y_{n+1} is determined solely from y_n. In this chapter, we consider multistep methods in which the computation of the numerical solution y_{n+1} uses the solution values at several previous nodes. We derive here two families of the most widely used multistep methods.

Reformulate the differential equation

$$Y'(t) = f(t, Y(t))$$

by integrating it over the interval $[t_n, t_{n+1}]$, obtaining

$$\int_{t_n}^{t_{n+1}} Y'(t)\, dt = \int_{t_n}^{t_{n+1}} f(t, Y(t))\, dt,$$

$$Y(t_{n+1}) = Y(t_n) + \int_{t_n}^{t_{n+1}} f(t, Y(t))\, dt. \tag{6.1}$$

We will develop numerical methods to compute the solution $Y(t)$ by approximating the integral in (6.1). There are many such methods, and we will consider only the most popular of them, the Adams–Bashforth (AB) and Adams–Moulton (AM) methods. These methods are the basis of some of the most widely used computer codes for solving the initial value problem. They are generally more efficient than the RK methods, especially if one wishes to find the solution with a high degree of accuracy or if the derivative function $f(t, y)$ is expensive to evaluate.

To evaluate the integral

$$\int_{t_n}^{t_{n+1}} g(t)\, dt, \qquad g(t) = Y'(t) = f(t, Y(t)), \tag{6.2}$$

we approximate $g(t)$ by using polynomial interpolation and then integrate the interpolating polynomial. For a given nonnegative integer q, the AB methods use interpolation polynomial of degree q at the points $\{t_n, t_{n-1}, \ldots, t_{n-q}\}$, and AM methods use interpolation polynomial of degree q at the points $\{t_{n+1}, t_n, t_{n-1}, \ldots, t_{n-q+1}\}$.

6.1 ADAMS–BASHFORTH METHODS

We begin with the AB method based on linear interpolation ($q = 1$). The linear polynomial interpolating $g(t)$ at $\{t_n, t_{n-1}\}$ is

$$p_1(t) = \frac{1}{h}[(t_n - t)g(t_{n-1}) + (t - t_{n-1})g(t_n)]. \tag{6.3}$$

From the theory of polynomial interpolation (Theorem B.3 in Appendix B),

$$g(t) - p_1(t) = \tfrac{1}{2}(t - t_n)(t - t_{n-1})g''(\zeta_n) \tag{6.4}$$

for some $t_{n-1} \le \zeta_n \le t_{n+1}$. Integrating over $[t_n, t_{n+1}]$, we obtain

$$\int_{t_n}^{t_{n+1}} g(t)\, dt \approx \int_{t_n}^{t_{n+1}} p_1(t)\, dt = \tfrac{1}{2}h[3g(t_n) - g(t_{n-1})].$$

In fact, we can obtain the more complete result

$$\int_{t_n}^{t_{n+1}} g(t)\, dt = \tfrac{1}{2}h[3g(t_n) - g(t_{n-1})] + \tfrac{5}{12}h^3 g''(\xi_n) \tag{6.5}$$

for some $t_{n-1} \le \xi_n \le t_{n+1}$; see Problem 4 for a derivation of a related but somewhat weaker result on the truncation error. Applying this to the relation (6.1) gives us

$$\begin{aligned} Y(t_{n+1}) = Y(t_n) + \tfrac{1}{2}h[3f(t_n, Y(t_n)) - f(t_{n-1}, Y(t_{n-1}))] \\ + \tfrac{5}{12}h^3 Y'''(\xi_n). \end{aligned} \tag{6.6}$$

Dropping the final term, the truncation error, we obtain the numerical method

$$y_{n+1} = y_n + \tfrac{1}{2}h[3f(t_n, y_n) - f(t_{n-1}, y_{n-1})]. \tag{6.7}$$

Table 6.1 An example of the second order Adams-Bashforth method

t	$y_h(t)$	$Y(t) - y_{2h}(t)$	$Y(t) - y_h(t)$	Ratio	$\frac{1}{3}[y_h(t) - y_{2h}(t)]$
2	0.49259722	2.13e $-$ 3	5.53e $-$ 4	3.9	5.26e $-$ 4
4	-1.41116963	2.98e $-$ 3	7.24e $-$ 4	4.1	7.52e $-$ 4
6	0.68174279	-3.91e $-$ 3	-9.88e $-$ 4	4.0	-9.73e $-$ 4
8	0.84373678	3.68e $-$ 4	1.21e $-$ 4	3.0	8.21e $-$ 5
10	-1.38398254	3.61e $-$ 3	8.90e $-$ 4	4.1	9.08e $-$ 4

With this method, note that it is necessary to have $n \geq 1$. Both y_0 and y_1 are needed in finding y_2, and y_1 cannot be found from (6.7). The value of y_1 must be obtained by another method. The method (6.7) is an example of a two step method, since values at t_{n-1} and t_n are needed in finding the value at t_{n+1}. If we assume $y_0 = Y_0$, and if we can determine $y_1 \approx Y(t_1)$ with an accuracy $\mathcal{O}(h^2)$, then the AB method (6.7) is of order 2, that is, its global error is of size $\mathcal{O}(h^2)$,

$$\max_{t_0 \leq t_n \leq b} |Y(t_n) - y_h(t_n)| \leq ch^2. \tag{6.8}$$

We must note that this result assumes $f(t, y)$ and $Y(t)$ are sufficiently differentiable, just as with all other similar convergence error bounds and asymptotic error results stated in this book. In this particular case (6.8), we would assume that $Y(t)$ is 3 times continuously differentiable on $[t_0, b]$ and that $f(t, y)$ satisfies the Lipschitz condition of (2.19) in Chapter 2. We usually omit the explicit statement as to the order of differentiability on $Y(t)$ being assumed, although it is usually apparent from the given error results.

Example 6.1 Use (6.7) to solve

$$Y'(t) = -Y(t) + 2\cos(t), \qquad Y(0) = 1 \tag{6.9}$$

with the solution $Y(t) = \sin(t) + \cos(t)$. For illustrative purposes only, we take $y_1 = Y(t_1)$. The numerical results are given in Table 6.1, using $h = 0.05$. Note that the errors decrease by a factor of approximately 4 when h is halved, which is consistent with the numerical method being of order 2. The Richardson error estimate is also included in the table, using the formula (5.13) for second-order methods. Where the error is decreasing like $\mathcal{O}(h^2)$, the error estimate is quite accurate. ∎

Adams methods are often considered to be "less expensive" than RK methods, and the main reason can be seen by comparing (6.7) with the second-order RK method in (5.20). The main task of both methods is to evaluate the derivative function $f(t, y)$. With second-order RK methods, there are two evaluations of f for each step from t_n to t_{n+1}. In contrast, the AB formula (6.7) uses only one evaluation per step, provided that past values of f are reused. Other factors affect the choice of a numerical method, but the AB and AM methods are generally more efficient in the number of evaluations of f that are needed for a given amount of accuracy.

A problem with multistep methods is the need to generate some of the initial values of the solution by using another method. For the second-order AB method in (6.7), we must obtain y_1; and since the global error in $y_h(t_n)$ is to be $\mathcal{O}(h^2)$, we must ensure that $Y(t_1) - y_h(t_1)$ is also $\mathcal{O}(h^2)$. There are two immediate possibilities, using methods from preceding chapters.

Case (1) Use Euler's method:

$$y_1 = y_0 + hf(t_0, y_0). \tag{6.10}$$

Assuming $y_0 = Y_0$, this has an error of

$$Y(t_1) - y_1 = \tfrac{1}{2}h^2 Y''(\xi_1)$$

based on (2.10) with $n = 0$. Thus (6.10) meets our error criteria for y_1. Globally, Euler's method has only $\mathcal{O}(h)$ accuracy, but the error of a single step is $\mathcal{O}(h^2)$.

Case (2) Use a second-order RK method, such as (5.20). Since only one step in t is being used, $Y(t_1) - y_1$ will be $\mathcal{O}(h^3)$, which is more than adequate.

Example 6.2 Combine (6.10) with (6.7) to solve the problem (6.9) from the last example. For $h = 0.05$ and $t = 10$, the error in the numerical solution turns out to be

$$Y(10) - y_h(10) \doteq 8.90 \times 10^{-4},$$

the same as before for the results in Table 6.1. ∎

Higher-order Adams–Bashforth methods are obtained by using higher degree polynomial interpolation in the approximation of the integrand in (6.2). (For an introduction to polynomial interpolation, see Appendix B.) The next higher-order example following the linear interpolation of (6.3) uses quadratic interpolation. Let $p_2(t)$ denote the quadratic polynomial that interpolates $g(t)$ at t_n, t_{n-1}, t_{n-2}, and then use

$$\int_{t_n}^{t_{n+1}} g(t)\, dt \approx \int_{t_n}^{t_{n+1}} p_2(t)\, dt.$$

To be more explicit, we may write

$$p_2(t) = g(t_n)\ell_0(t) + g(t_{n-1})\ell_1(t) + g(t_{n-2})\ell_2(t) \tag{6.11}$$

with

$$
\left.
\begin{aligned}
\ell_0(t) &= \frac{(t - t_{n-1})(t - t_{n-2})}{2h^2}, \\[2mm]
\ell_1(t) &= -\frac{(t - t_n)(t - t_{n-2})}{h^2}, \\[2mm]
\ell_2(t) &= \frac{(t - t_n)(t - t_{n-1})}{2h^2}.
\end{aligned}
\right\} \tag{6.12}
$$

For the error, we have

$$g(t) - p_2(t) = \tfrac{1}{6}(t - t_n)(t - t_{n-1})(t - t_{n-2})g'''(\zeta_n) \qquad (6.13)$$

for some $t_{n-2} \le \zeta_n \le t_{n+1}$.

It can be shown that

$$\int_{t_n}^{t_{n+1}} g(t)\,dt = \tfrac{1}{12}h[23g(t_n) - 16g(t_{n-1}) + 5g(t_{n-2})] + \tfrac{3}{8}h^4 g'''(\xi_n)$$

for some $t_{n-2} \le \xi_n \le t_{n+1}$. Applying this to (6.1), the integral formulation of the differential equation, we obtain

$$Y(t_{n+1}) = Y(t_n) + \tfrac{1}{12}h[23f(t_n, Y(t_n)) - 16f(t_{n-1}, Y(t_{n-1}))$$
$$+ 5f(t_{n-2}, Y(t_{n-2}))] + \tfrac{3}{8}h^4 Y^{(4)}(\xi_n).$$

By dropping the last term, the truncation error, we obtain the third-order AB method

$$y_{n+1} = y_n + \tfrac{1}{12}h[23y'_n - 16y'_{n-1} + 5y'_{n-2}], \qquad n \ge 2, \qquad (6.14)$$

where $y'_k \equiv f(t_k, y_k)$, $k \ge 0$. This is a three step method, requiring $n \ge 2$. Thus y_1, y_2 must be obtained separately by other methods. We leave the implementation and illustration of (6.14) as Problem 2 for the reader.

In general, it can be shown that the AB method based on interpolation of degree q will be a $(q+1)$-step method, and its truncation error will be of the form

$$T_{n+1} = c_q h^{q+2} Y^{(q+2)}(\xi_n)$$

for some $t_{n-q} \le \xi_n \le t_{n+1}$. The initial values $y_1, \ldots, y_q$ will have to be generated by other methods. If the errors in these initial values satisfy

$$Y(t_n) - y_h(t_n) = \mathcal{O}(h^{q+1}), \qquad n = 1, 2, \ldots, q, \qquad (6.15)$$

then the global error in the $(q+1)$-step AB method will also be $\mathcal{O}(h^{q+1})$, provided that the true solution Y is sufficiently differentiable. In addition, the global error will satisfy an asymptotic error formula

$$Y(t_n) - y_h(t_n) = D(t_n)h^{q+1} + \mathcal{O}(h^{q+2}),$$

much as was true earlier for the Taylor and RK methods described in Chapter 5. Thus Richardson's extrapolation can be used to accelerate the convergence of the method and to estimate the error.

To generate the initial values $y_1, \ldots, y_q$ for the $(q+1)$-step AB method, and to have their errors satisfy the requirement (6.15), it is sufficient to use a RK method of order q. However, in many instances, people prefer to use a RK method of order $q+1$, the same order as that of the $(q+1)$-step AB method. Other procedures are used in the automatic computer programs for AB methods, and we discuss them later in this chapter.

Table 6.2   Adams-Bashforth methods

q	Order	Method	T. Error
0	1	$y_{n+1} = y_n + hy'_n$	$\frac{1}{2}h^2 Y''(\xi_n)$
1	2	$y_{n+1} = y_n + \frac{h}{2}[3y'_n - y'_{n-1}]$	$\frac{5}{12}h^3 Y'''(\xi_n)$
2	3	$y_{n+1} = y_n + \frac{h}{12}[23y'_n - 16y'_{n-1} + 5y'_{n-2}]$	$\frac{3}{8}h^4 Y^{(4)}(\xi_n)$
3	4	$y_{n+1} = y_n + \frac{h}{24}[55y'_n - 59y'_{n-1} + 37y'_{n-2} - 9y'_{n-3}]$	$\frac{251}{720}h^5 Y^{(5)}(\xi_n)$

Table 6.3   Example of fourth order Adams-Bashforth method

t	$y_h(t)$	$Y(t) - y_{2h}(t)$	$Y(t) - y_h(t)$	Ratio	$\frac{1}{15}[y_h(t) - y_{2h}(t)]$
2	0.49318680	$-3.96e - 4$	$-3.62e - 5$	10.9	$-2.25e - 5$
4	1.41037698	$-1.25e - 3$	$-6.91e - 5$	18.1	$-7.37e - 5$
6	0.68067962	$1.05e - 3$	$7.52e - 5$	14.0	$6.12e - 5$
8	0.84385416	$3.26e - 4$	$4.06e - 6$	80.0	$2.01e - 5$
10	-1.38301376	$-1.33e - 3$	$-7.89e - 5$	16.9	$-7.82e - 5$

The AB methods of orders 1 through 4 are given in Table 6.2; the column heading "T. Error" denotes "Truncation Error". The order 1 formula is simply Euler's method. In the table, $y'_k \equiv f(t_k, y_k)$.

Example 6.3 Solve the problem (6.9) by using the fourth-order AB method. Since we are illustrating the AB method, we simply generate the initial values y_1, y_2, y_3 by using the true solution,

$$y_i = Y(t_i), \quad i = 1, 2, 3.$$

The results for $h = 0.125$ and $2h = 0.25$ are given in Table 6.3. Richardson's error estimate for a fourth-order method is given in the last column. For a fourth-order method, the error should decrease by a factor of approximately 16 when h is halved. In those cases where this is true, the Richardson's error estimate is accurate. In no case is the error badly underestimated. ∎

Comparing these results with those in Table 5.6 for the fourth-order Fehlberg method, we see that the present errors appear to be very large. But note that the Fehlberg formula uses five evaluations of $f(t, y)$ for each step of t_n to t_{n+1}; whereas the fourth-order AB method uses only one evaluation of f per step, assuming that previous evaluations are reused. If this AB method is used with an h that is only $\frac{1}{5}$ as large (for a comparable number of evaluations of f), then the present errors will decrease by a factor of approximately $5^4 = 625$. The AB errors will be mostly smaller than those of the Fehlberg method in Table 5.6, and the work will be comparable (measured by the number of evaluations of f).

Table 6.4   Example of Adams-Moulton method of order 2

t	$Y(t) - y_{2h}(t)$	$Y(t) - y_h(t)$	Ratio	$\frac{1}{3}[y_h(t) - y_{2h}(t)]$
2	$-4.59e - 4$	$-1.15e - 4$	4.0	$-1.15e - 4$
4	$-5.61e - 4$	$-1.40e - 4$	4.0	$-1.40e - 4$
6	$7.98e - 4$	$2.00e - 4$	4.0	$2.00e - 4$
8	$-1.21e - 4$	$-3.04e - 5$	4.0	$-3.03e - 4$
10	$-7.00e - 4$	$-1.75e - 4$	4.0	$-1.28e - 4$

6.2 ADAMS–MOULTON METHODS

As with the AB methods, we begin our presentation of AM methods by considering the method based on linear interpolation. Let $p_1(t)$ be the linear polynomial that interpolates $g(t)$ at t_n and t_{n+1},

$$p_1(t) = \frac{1}{h}[(t_{n+1} - t)g(t_n) + (t - t_n)g(t_{n+1})].$$

Using this equation to approximate the integrand in (6.2), we obtain the trapezoidal rule discussed in Chapter 4,

$$Y(t_{n+1}) = Y(t_n) + \tfrac{1}{2}h[f(t_n, Y(t_n)) + f(t_{n+1}, Y(t_{n+1}))] - \tfrac{1}{12}h^3 Y'''(\xi_n). \quad (6.16)$$

Dropping the last term, the truncation error, we obtain the AM method

$$y_{n+1} = y_n + \tfrac{1}{2}h[f(t_n, y_n) + f(t_{n+1}, y_{n+1})], \qquad n \geq 0. \quad (6.17)$$

This is the trapezoidal method discussed in Section 4.2. It is a second-order method and has a global error of size $\mathcal{O}(h^2)$. Moreover, it is absolutely stable.

Example 6.4 Solve the earlier problem (6.9) by using the AM method (6.17) (the trapezoidal method). The results are given in Table 6.4 for $h = 0.05, 2h = 0.1$, and the Richardson error estimate for second-order methods is given in the last column. In this case, the $\mathcal{O}(h^2)$ error behavior is very apparent, and the error estimation is very accurate. ∎

Example 6.5 Repeat Example 6.4, but using the procedure described following (4.28) in Chapter 4, with only one iterate being computed for each n. Then, the errors do not change significantly from those given in Table 6.4. For example, with $t = 10$ and $h = 0.05$, the error is

$$Y(10) - y_h(10) \doteq -2.02 \times 10^{-4}.$$

This is not very different from the value of -1.75×10^{-4} given in Table 6.4. The use of the iterate $y_{n+1}^{(1)}$ as the root y_{n+1} will not affect significantly the accuracy of

Table 6.5 Adams-Moulton methods

q	Order	Method	T. Error
0	1	$y_{n+1} = y_n + h y'_{n+1}$	$-\frac{1}{2} h^2 Y''(\xi_n)$
1	2	$y_{n+1} = y_n + \frac{h}{2}[y'_{n+1} + y'_n]$	$-\frac{1}{12} h^3 Y'''(\xi_n)$
2	3	$y_{n+1} = y_n + \frac{h}{12}[5y'_{n+1} + 8y'_n - y'_{n-1}]$	$-\frac{1}{24} h^4 Y^{(4)}(\xi_n)$
3	4	$y_{n+1} = y_n + \frac{h}{24}[9y'_{n+1} + 19y'_n - 5y'_{n-1} + y'_{n-2}]$	$-\frac{19}{720} h^5 Y^{(5)}(\xi_n)$

the solution for most differential equations. Stiff differential equations are a major exception. ∎

By integrating the polynomial of degree q that interpolates on the set of the nodes $\{t_{n+1}, t_n, \ldots, t_{n-q+1}\}$ to the function $g(t)$ of (6.2), we obtain the AM method of order $q + 1$. It will be an implicit method, but in other respects the theory is the same as for the AB methods described previously. The AM methods of orders 1 through 4 are given in Table 6.5, where $y'_k \equiv f(t_k, y_k)$. As in Table 6.2, the column heading "T. Error" denotes "Truncation Error". Note that the AM method of order 1 is the backward Euler method, and the AM method of order 2 is the trapezoidal method.

The effective cost of an AM method is two evaluations of the derivative $f(t, y)$ per step in most cases and assuming that previous function values of f are reused. This includes one evaluation of f to calculate an initial guess $y_{n+1}^{(0)}$, and then one evaluation of f in the iteration formula for the AM method. For example, with the trapezoidal method this means using the calculation

$$
\begin{aligned}
y_{n+1}^{(0)} &= y_n + \tfrac{1}{2}h\left[3f(t_n, y_n) - f(t_{n-1}, y_{n-1})\right], \\
y_{n+1}^{(1)} &= y_n + \tfrac{1}{2}h[f(t_n, y_n) + f(t_{n+1}, y_{n+1}^{(0)})],
\end{aligned}
\tag{6.18}
$$

or using some other *predictor* formula for $y_{n+1}^{(0)}$ with an equivalent accuracy. With this calculation, there is no significant gain in accuracy over the AB method of the same order when comparing methods of equivalent cost.

Nonetheless, AM methods possess other properties that make them desirable for use in many types of differential equations. The desirable features relate to stability characteristics of numerical methods. Recall from Chapter 4, following (4.3), that we study the behavior of a numerical method when applied to the model problem

$$
\begin{aligned}
Y'(t) &= \lambda Y(t), \quad t > 0, \\
Y(0) &= 1.
\end{aligned}
\tag{6.19}
$$

We always assume the constant $\lambda < 0$ or λ is complex with $\text{Real}(\lambda) < 0$. The true solution of the problem (6.19) is $Y(t) = e^{\lambda t}$, which decays exponentially in t since the parameter λ has a negative real part. The kind of stability property that we would like for a numerical method is that when it is applied to (6.19), the numerical solution satisfies

$$y_h(t_n) \to 0 \quad \text{as} \quad t_n \to \infty \tag{6.20}$$

for any choice of stepsize h. With most numerical methods, this is not satisfied. The set of values $h\lambda$, considered as a subset of the complex plane, for which $y_n \to 0$ as $n \to \infty$, is called the *region of absolute stability* of the numerical method.

As seen in Chapter 4, the AM methods of orders 1 and 2 are absolutely stable, satisfying (6.20) for all values of h. Such methods are particularly suitable for solving stiff differential equations. In general, we prefer numerical methods with a larger region of absolute stability; the larger is the region, the less restrictive the condition on h in order to ensure satisfaction of (6.20) for the model problem (6.19). Thus a method with a large region of absolute stability is generally preferred over a method with a smaller region, provided that the accuracy of the two methods is similar. It can be shown that for AB and AM methods of equal order, the AM method will have the larger region of absolute stability; see Figures 8.1 and 8.2 in Chapter 8. Consequently, Adams–Moulton methods are generally preferred over Adams–Bashforth methods.

Example 6.6 Applying the AB method of order 2 to equation (6.19) leads to the finite difference equation

$$y_{n+1} = y_n + \tfrac{1}{2}h\lambda\left(3y_n - y_{n-1}\right), \quad n = 1, 2, \ldots \tag{6.21}$$

with y_0 and y_1 determined beforehand. Jumping ahead to (7.45) in Chapter 7, the solution to this finite difference equation is given by

$$y_h(t_n) = \gamma_0\left[r_0(h\lambda)\right]^n + \gamma_1\left[r_1(h\lambda)\right]^n, \quad n \geq 0 \tag{6.22}$$

with $r_0(h\lambda)$ and $r_1(h\lambda)$ the roots of the quadratic polynomial

$$r^2 = r + \tfrac{1}{2}h\lambda\left(3r - 1\right). \tag{6.23}$$

When $\lambda = 0$, one of the roots equals 1, and we denote arbitrarily that root by $r_0(h\lambda)$ in general: $r_0(0) = 1$. The constants γ_0 and γ_1 are determined from y_0 and y_1. In order to satisfy (6.22) for a given choice of $h\lambda$ and for any choice of γ_0 and γ_1, it is necessary to have

$$|r_0(h\lambda)| < 1, \quad |r_1(h\lambda)| < 1. \tag{6.24}$$

Solving this pair of inequalities for the case that λ is real, and looking only at the case that $\lambda < 0$, we obtain

$$-1 < h\lambda < 0 \tag{6.25}$$

as the region of absolute stability on the real axis. In contrast, the AM method of order 2 has $-\infty < h\lambda < 0$ on the real axis of its region of stability. There is no stability restriction on h with this AM method. ∎

6.3 COMPUTER CODES

Some of the most popular computer codes for solving the initial value problem are based on using AM and AB methods in combination, as suggested in the discussion preceding (6.18). These codes control the truncation error by varying both the stepsize h and the order of the method. They are self-starting in terms of generating the initial values $y_1, \ldots, y_q$ needed with higher-order methods of order $q+1$. To generate these values, they begin with first-order methods and a small stepsize h and then increase the order to generate the starting values needed with higher-order methods. The possible order is allowed to be as large as 12 or more; this results in a very efficient numerical method when the solution $Y(t)$ has several continuous derivatives and is slowly varying. A comprehensive discussion of Adams' methods and an example of one such computer code is given in Shampine [72].

MATLAB® program. To facilitate the illustrative programming of the methods of this chapter, we present a modification of the Euler program of Chapter 2. The program implements the Adams–Bashforth formula of order 2, given in (6.7); and it uses Euler's method to generate the first value y_1 as in (6.10). We defer to the Problems section the experimental use of this program.

```
function [t,y] = AB2(t0,y0,t_end,h,fcn)
%
% function [t,y]=AB2(t0,y0,t_end,h,fcn)
%
% Solve the initial value problem
%    y' = f(t,y),   t0 <= t <= b,   y(t0)=y0
% Use Adams-Bashforth formula of order 2 with
% a stepsize of h. Euler's method is used for
% the value y1.  The user must supply a program for
% the right side function defining the differential
% equation.   For some name, say deriv, use a first
% line of the form
%    function ans=deriv(t,y)
% A sample call would be
%    [t,z]=AB2(t0,z0,b,delta,'deriv')
%
% Output:
% The routine AB2 will return two vectors, t and y.
% The vector t will contain the node points
%    t(1)=t0, t(j)=t0+(j-1)*h, j=1,2,...,N
% with
%    t(N) <= t_end-h,   t(N)+h > t_end-h
% The vector y will contain the estimates of the
% solution Y at the node points in t.
%
n = fix((t_end-t0)/h)+1;
```

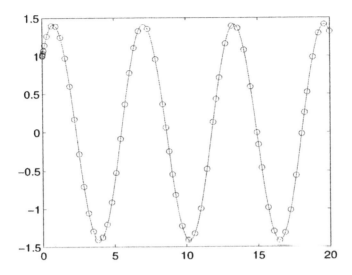

Figure 6.1 The solution values to (6.26) obtained by ode113 are indicated by the symbol o. The curve line is obtained by interpolating these solution values from ode113 using deval.

```
t = linspace(t0,t0+(n-1)*h,n)';
y = zeros(n,1);
y(1) = y0;
ft1 = feval(fcn,t(1),y(1));
y(2) = y(1)+h*ft1;
for i = 3:n
  ft2 = feval(fcn,t(i-1),y(i-1));
  y(i) = y(i-1)+h*(3*ft2-ft1)/2;
  ft1 = ft2;
end
```

6.3.1 MATLAB ODE codes

Built-in MATLAB programs based on multistep methods are ode113 and ode15s. These programs implement explicit and implicit linear multistep methods of various orders, respectively. The program ode113 is used to solve nonstiff ordinary differential equations, using the Adams–Bashforth and Adams–Moulton methods presented in this chapter. The code ode15s is for stiff ordinary differential equations, and it is based on yet another variable order family of multistep methods, one that is discussed in Chapter 8. The programs are used in precisely the same manner as the program ode45 discussed in Section 5.5 of Chapter 5; and the entire suite of MAT-

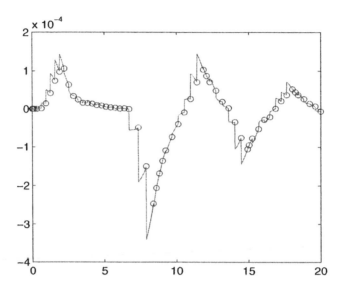

Figure 6.2   The errors in the solution to (6.26) obtained using ode113.   The errors at the node points are indicated by the symbol o

LAB ode programs is discussed at length by Shampine and Reichelt [73]. Also, see Shampine [72] for a thorough study of one-step and multistep methods and of their implementation in computer software.

Example 6.7 We modify the program test_ode45 by replacing ode45 with ode113 throughout the code. The program ode113 is recommended for medium- to high-accuracy solutions, but we will illustrate its use with the same example as in Section 5.5 of Chapter 5 for the program ode45. As before, we solve the test equation

$$Y'(t) = -Y(t) + 2\cos(t), \qquad Y(0) = 1 \tag{6.26}$$

and we use $AbsTol = 10^{-6}$, $RelTol = 10^{-4}$. Figures 6.1 and 6.2 illustrate, respectively, the interpolated numerical solution and the error contained therein. Compare these results to those in Figures 5.2 and 5.3 of Chapter 5. There are 229 derivative evaluations when using ode45 for this problem, whereas ode113 uses 132 evaluations. This is a typical example for comparison of the number of derivative evaluations. ■

PROBLEMS

1. Use the MATLAB program for the AB method of order two to solve the equations in Problem 2 of Chapter 5. Include the Richardson error estimate for $y_h(t)$ when $h = 0.1$ and 0.05.

2. Modify the MATLAB program of this chapter to use the third-order AB method. To calculate y_1 and y_2, use one of the second-order RK methods from Chapter 5. Then repeat Problem 1.

3. Use the program from Problem 2 to solve the continuing example problem (6.9).

4. To make the error term in (6.5) a bit more believable, prove

$$\int_0^h \gamma(s)\, ds - \tfrac{1}{2} h[3\gamma(0) - \gamma(-h)] = \tfrac{5}{12} h^3 \gamma''(0) + \mathcal{O}(h^4)$$

with $\gamma(s)$ a 3 times continuously differentiable function for $-h \le s \le h$.
Hint: Expand $\gamma(s)$ as a quadratic Taylor polynomial about the origin, with an error term $R_3(t)$. Substitute that Taylor expansion into the left side of the equation above, and obtain the right side. For simplicity, we have changed the interval in (6.5) from $[t_n, t_{n+1}]$ to $[0, h]$. The result extends to (6.5) by means of a simple change of variable in (6.5), namely, $t = t_n + s, 0 \le s \le h$. Also note that if $-h \le \xi \le h$, then

$$\gamma''(\xi) = \gamma''(0) + \xi \gamma'''(\zeta), \qquad \text{some } \zeta \text{ between } 0 \text{ and } \xi$$
$$= \gamma''(0) + \mathcal{O}(h),$$
$$\tfrac{5}{12} h^3 \gamma''(\xi) = \tfrac{5}{12} h^3 \gamma''(0) + \mathcal{O}(h^4),$$

since $|\xi| \le h$. This argument assumes $\gamma(s)$ is 3 times continuously differentiable.

5. Repeat the type of argument given in Problem 4, extending it to the Adams–Bashforth method of order 3, given in Table 6.2.

6. Repeat the type of argument given in Problem 4, extending it to the Adams–Moulton method of order 2, given in Table 6.5.

7. Repeat the type of argument given in Problem 4, extending it to the Adams–Moulton method of order 3, given in Table 6.5.

8. Modify the MATLAB program of this chapter to use the AM method of order 2. For the predictor, use the AB method of order 2; for the first step y_1, use the Euler predictor. Iterate the formula (4.25) only once. Apply this to the solution of the equations considered in Problem 1, and produce the Richardson error estimate.

9. Use the MATLAB code ode113 to solve the equations in Problem 2 of Chapter 5. For error tolerances, use absolute error bounds $AbsTol = 10^{-4}$ and $\epsilon = 10^{-6}$, along with a relative error tolerance $RelTol = 10^{-8}$. Keep track of the number of evaluations of $f(t, y)$ that are used by the routine, and compare it to the number used in your own programs for the Adams–Bashforth and Adams–Moulton methods.

10. (a) Using the program of Problem 1 for the AB method of order 2, solve

$$Y'(t) = -50Y(t) + 51\cos(t) + 49\sin(t), \qquad Y(0) = 1$$

for $0 \le t \le 10$. The solution is $Y(t) = \sin(t) + \cos(t)$. Use stepsizes of $h = 0.1, 0.02, 0.01$. In each case, print the errors as well as the answers.

   (b) Using the program of Problem 8 for the AM method of order 2, repeat part (a). Check the condition of (4.26).

   (c) When the AM method of order 2 is applied to the equation in (a), the value of y_{n+1} can be found directly. While doing so, repeat part (a). Compare your results.

11. The Adams–Bashforth and Adams–Moulton methods are based on (6.1) together with the integration over $[t_n, t_{n+1}]$ of a polynomial interpolating the integrand $Y'(t) = f(t, Y(t))$. As an alternative, consider integration over $[t_{n-1}, t_{n+1}]$, obtaining

$$Y(t_{n+1}) = Y(t_{n-1}) + \int_{t_{n-1}}^{t_{n+1}} f(t, Y(t))\, dt. \tag{6.27}$$

We can replace the integrand $f(t, Y(t))$ with an approximation based on interpolation. The simplest example is to use a constant interpolant; in particular,

$$\int_{t_{n-1}}^{t_{n+1}} f(t, Y(t))\, dt \approx \int_{t_{n-1}}^{t_{n+1}} f(t_n, Y(t_n))\, dt = 2hf(t_n, Y(t_n)).$$

This leads to the numerical method

$$y_{n+1} = y_{n-1} + 2hf(t_n, y_n), \qquad n \ge 1. \tag{6.28}$$

This is called the *midpoint method*. As with the Adams–Bashforth method (6.7) of order 2, the value of y_1 must be obtained by other means. Using the type of argument given in Problem 4, show that

$$Y(t_{n+1}) - [Y(t_{n-1}) + 2hf(t_n, Y(t_n))] = -\tfrac{1}{3}h^3 Y'''(t_n) + \mathcal{O}(h^4).$$

Hint: Expand $Y(t)$ as a quadratic Taylor polynomial about t_n, with an error term $R_3(t)$. Substitute that Taylor expansion into the left side of the equation above to obtain the right side.

12. Using the same arguments as in Problem 11, consider interpolating $Y'(t) = f(t, Y(t))$ with a quadratic polynomial. Have it interpolate $Y'(t) = f(t, Y(t))$ at the nodes $\{t_{n-1}, t_n, t_{n+1}\}$. Use this to obtain the numerical method

$$\begin{aligned} y_{n+1} = y_{n-1} + \tfrac{1}{3}[hf(t_{n-1}, y_{n-1}) \\ + 4f(t_n, y_n) + f(t_{n+1}, y_{n+1})]. \end{aligned} \tag{6.29}$$

As with the Adams–Moulton methods, this is an implicit method and the value of y_{n+1} must be calculated by a rootfinding method. Also, the value of y_1 must be obtained by other means.

This is *Simpson's parabolic rule* for numerical integration, and when applied, as here, to solving differential equations, it is one part of *Milne's method*, which is mainly of historical interest, as the family of Adams methods have replaced it in modern codes. We return to Simpson's rule, however, when developing numerical methods for solving Volterra integral equations in Chapter 12.

13. As an alternative to (6.27), consider

$$Y(t_{n+1}) = Y(t_{n-3}) + \int_{t_{n-3}}^{t_{n+1}} f(t, Y(t))\, dt.$$

Using the same arguments as in Problem 12, consider interpolating $Y'(t) = f(t, Y(t))$ with a quadratic polynomial, but have it interpolate $Y'(t)$ at the nodes $\{t_{n-2}, t_{n-1}, t_n\}$. Use this to obtain the numerical method

$$
\begin{aligned}
y_{n+1} = y_{n-3} + \tfrac{4}{3}h[2f\,(t_{n-2}, y_{n-2}) \\
- f\,(t_{n-1}, y_{n-1}) + 2f\,(t_n, y_n)].
\end{aligned}
\tag{6.30}
$$

This is an explicit method, and historically it has been used to estimate an initial value $y_{n+1}^{(0)}$ for the iterative solution of equation (6.29) in Problem 12, thus forming the other half of *Milne's method*. The values of y_1, y_2, y_3 must be obtained by other means.

14. Repeat Problems 20 and 21 of Chapter 5 using the MATLAB code ode113. Do not attempt to estimate the error since that is embedded in ode113.

15. Repeat Problem 24 of Chapter 5 using the MATLAB code ode113.

CHAPTER 7

GENERAL ERROR ANALYSIS FOR MULTISTEP METHODS

We now present a general error analysis for multistep methods in solving the initial value problem of a single first-order equation. In addition to explaining the underlying behavior of the numerical methods, such a general error analysis allows us to design better numerical procedures for various classes of problems. We begin by considering the truncation error for multistep methods. Next, in Section 7.2, we look at a relatively simple error analysis that is similar to that given for Euler's method in Chapter 2; it is an error analysis that works for many popular multistep methods. In Section 7.3 we give a complete error analysis for all multistep methods, and we follow it with some examples.

As before, let $h > 0$ and define the nodes by $t_n = t_0 + nh$, $n \geq 0$. The general form of the multistep methods to be considered is

$$y_{n+1} = \sum_{j=0}^{p} a_j y_{n-j} + h \sum_{j=-1}^{p} b_j f(t_{n-j}, y_{n-j}), \qquad n \geq p. \tag{7.1}$$

The coefficients $a_0, \ldots, a_p$, b_{-1}, $b_0, \ldots, b_p$ are constants and $p \geq 0$. Assuming that $|a_p| + |b_p| \neq 0$, we consider this method a $(p+1)$-step method, because $p+1$ previous solution values are being used to compute y_{n+1}. The values $y_1, \ldots, y_p$ must

111

be obtained by other means, as was illustrated in Chapter 6 with the Adams methods. Euler's method is an example of a one-step method with $p = 0$ and

$$a_0 = 1, \qquad b_0 = 1, \qquad b_{-1} = 0.$$

If $b_{-1} = 0$, then y_{n+1} occurs on only the left side of equation (7.1). Such formulas are called *explicit methods*. If $b_{-1} \neq 0$, then y_{n+1} is present on both sides of (7.1), and the formula is called an *implicit method*. As was discussed following (4.12) in Chapter 4 for the backward Euler method, the solution y_{n+1} can be computed by fixed point iteration,

$$y_{n+1}^{(i+1)} = \sum_{j=0}^{p} a_j y_{n-j} + h \sum_{j=0}^{p} b_j f(t_{n-j}, y_{n-j}) + h b_{-1} f(t_{n+1}, y_{n+1}^{(i)}), \quad i = 0, 1, \ldots,$$

provided h is chosen sufficiently small.

Example 7.1

1. The midpoint method is defined by

$$y_{n+1} = y_{n-1} + 2hf(t_n, y_n), \qquad n \geq 1 \tag{7.2}$$

   and it is an explicit two-step method. We discuss this method in more detail later in the chapter.

2. The Adams–Bashforth and Adams–Moulton methods are all special cases of (7.1), with

$$a_0 = 1, \qquad a_j = 0 \quad \text{for} \quad j = 1, \ldots, p.$$

   Also, refer to the formulas for these methods in Tables 6.2 and 6.5 of Chapter 6. ∎

7.1 TRUNCATION ERROR

For any differentiable function $Y(t)$, define the truncation error for integrating $Y'(t)$ by

$$T_n(Y) = Y(t_{n+1}) - \left[\sum_{j=0}^{p} a_j Y(t_{n-j}) + h \sum_{j=-1}^{p} b_j Y'(t_{n-j}) \right] \tag{7.3}$$

for $n \geq p$. Define the function $\tau_n(Y)$ by

$$\tau_n(Y) = \frac{1}{h} T_n(Y). \tag{7.4}$$

In order to prove the convergence of the approximate solution $\{y_n : t_0 \leq t_n \leq b\}$ of (7.1) to the solution $Y(t)$ of the initial value problem

$$Y'(t) = f(t, Y(t)), \quad t \geq t_0,$$
$$Y(t_0) = Y_0,$$

it is necessary to have

$$\tau(h) \equiv \max_{t_p \le t_n \le b} |T_n(Y)| \to 0 \qquad \text{as } h \to 0. \tag{7.5}$$

This is often called the *consistency condition* for the method (7.1). The speed of convergence of the solution $\{y_n\}$ to the true solution $Y(t)$ is related to the speed of convergence in (7.5), and thus we need to know the conditions under which

$$\tau(h) = \mathcal{O}(h^m) \tag{7.6}$$

for some desired choice of $m \ge 1$. We now examine the implications of (7.5) and (7.6) for the coefficients in (7.1).

Theorem 7.2 *Let $m \ge 1$ be a given integer. For (7.5) to hold for all continuously differentiable functions $Y(t)$, that is, for the method (7.1) to be consistent, it is necessary and sufficient that*

$$\sum_{j=0}^{p} a_j = 1, \tag{7.7}$$

$$-\sum_{j=0}^{p} j a_j + \sum_{j=-1}^{p} b_j = 1. \tag{7.8}$$

Further, for (7.6) to be valid for all functions $Y(t)$ that are $m+1$ times continuously differentiable, it is necessary and sufficient that (7.7)–(7.8) hold and that

$$\sum_{j=0}^{p} (-j)^i a_j + i \sum_{j=-1}^{p} (-j)^{i-1} b_j = 1, \qquad i = 2, \ldots, m. \tag{7.9}$$

Proof. Note that

$$T_n(\alpha Y + \beta W) = \alpha T_n(Y) + \beta T_n(W) \tag{7.10}$$

for all constants α, β and all differentiable functions Y, W. To examine the consequences of (7.5) and (7.6), expand $Y(t)$ about t_n using Taylor's theorem to obtain

$$Y(t) = \sum_{i=0}^{m} \frac{1}{i!} (t - t_n)^i Y^{(i)}(t_n) + R_{m+1}(t), \tag{7.11}$$

$$R_{m+1}(t) = \frac{1}{m!} \int_{t_n}^{t} (t - s)^m Y^{(m+1)}(s)\, ds$$

$$= \frac{(t - t_n)^{m+1}}{(m+1)!} Y^{(m+1)}(\xi_n) \tag{7.12}$$

with ξ_n between t and t_n (see (A.4)–(A.6) in Appendix A). We are assuming that $Y(t)$ is $m+1$ times continuously differentiable on the interval bounded by t and t_n. Substituting into (7.3) and using (7.10), we obtain

$$T_n(Y) = \sum_{i=0}^{m} \frac{1}{i!} Y^{(i)}(t_n) T_n((t - t_n)^i) + T_n(R_{m+1}).$$

It is necessary to calculate $T_n((t - t_n)^i)$ for $i \geq 0$.

- For $i = 0$,

$$T_n(1) = c_0 \equiv 1 - \sum_{j=0}^{p} a_j.$$

- For $i \geq 1$,

$$T_n((t - t_n)^i) = (t_{n+1} - t_n)^i$$

$$- \left[\sum_{j=0}^{p} a_j(t_{n-j} - t_n)^i + h \sum_{j=-1}^{p} b_j i(t_{n-j} - t_n)^{i-1} \right]$$

$$= c_i h^i$$

$$c_i = 1 - \left[\sum_{j=0}^{p} (-j)^i a_j + i \sum_{j=-1}^{p} (-j)^{i-1} b_j \right] \qquad i \geq 1. \quad (7.13)$$

This gives

$$T_n(Y) = \sum_{i=0}^{m} \frac{c_i}{i!} h^i Y^{(i)}(t_n) + T_n(R_{m+1}). \tag{7.14}$$

From (7.12) it is straightforward that $T_n(R_{m+1}) = \mathcal{O}(h^{m+1})$. If Y is $m + 2$ times continuously differentiable, we may write the remainder $R_{m+1}(t)$ as

$$R_{m+1}(t) = \frac{1}{(m + 1)!}(t - t_n)^{m+1} Y^{(m+1)}(t_n) + \cdots,$$

and then

$$T_n(R_{m+1}) = \frac{c_{m+1}}{(m + 1)!} h^{m+1} Y^{(m+1)}(t_n) + \mathcal{O}(h^{m+2}). \tag{7.15}$$

To obtain the consistency condition (7.5), assuming that Y is an arbitrary twice continuously differentiable function, we need $\tau(h) = \mathcal{O}(h)$ and this requires $T_n(Y) = \mathcal{O}(h^2)$. Using (7.14) with $m = 1$, we must have $c_0 = c_1 = 0$, which gives the set of equations (7.7)–(7.8). In some texts, these equations are referred to as the *consistency conditions*. It can be further shown that (7.7)–(7.8) are the necessary and sufficient conditions for the consistency (7.5), even when Y is only assumed to be continuously differentiable. To obtain (7.6) for some $m \geq 1$, we must have $T_n(Y) = \mathcal{O}(h^{m+1})$. From (7.14) and (7.13), this will be true if and only if $c_i = 0$, $i = 0, 1, \ldots, m$. This proves the conditions (7.9) and completes the proof. ∎

The largest value of m for which (7.6) holds is called the *order* or *order of convergence* of the method (7.1).

Example 7.3 Find all second-order two-step methods. Formula (7.1) is

$$y_{n+1} = a_0 y_n + a_1 y_{n-1} + h \left[b_{-1} f(t_{n+1}, y_{n+1}) + b_0 f(t_n, y_n) \right. \tag{7.16}$$
$$\left. + b_1 f(t_{n-1}, y_{n-1}) \right], \qquad n \geq 1.$$

The coefficients must satisfy (7.7)–(7.9) with $m = 2$:

$$a_0 + a_1 = 1, \qquad -a_1 + b_{-1} + b_0 + b_1 = 1, \qquad a_1 + 2b_{-1} - 2b_1 = 1.$$

Solving, we obtain

$$a_1 = 1 - a_0, \qquad b_{-1} = 1 - \tfrac{1}{4}a_0 - \tfrac{1}{2}b_0, \qquad b_1 = 1 - \tfrac{3}{4}a_0 - \tfrac{1}{2}b_0 \tag{7.17}$$

with a_0, b_0 indeterminate. The midpoint method is a special case in which $a_0 = 0$, $b_0 = 2$. For the truncation error, we have

$$T_n(R_3) = \tfrac{1}{6}c_3 h^3 Y^{(3)}(t_n) + \mathcal{O}(h^4), \tag{7.18}$$
$$c_3 = -4 + 2a_0 + 3b_0. \tag{7.19}$$

The coefficients a_0, b_0 can be chosen to improve the stability, give a small truncation error, give an explicit formula, or some combination of these. The conditions to ensure stability and convergence cannot be identified until the general theory for (7.1) has been given in the remainder of this chapter. ■

7.2 CONVERGENCE

We now give a convergence result for the numerical method (7.1). Although the theorem will not cover all the multistep methods that are convergent, it does include many methods of current interest, including those of Chapters 2, 4, and 6. Moreover, the proof is much easier than that of the more general Theorem 7.6 given in Section 7.3.

Theorem 7.4 *Consider solving the initial value problem*

$$Y'(t) = f(t, Y(t)), \quad t \geq t_0, \tag{7.20}$$
$$Y(t_0) = Y_0$$

using the multistep method (7.1). Assume that the derivative function $f(t, y)$ is continuous and satisfies the Lipschitz condition

$$|f(t, y_1) - f(t, y_2)| \leq K |y_1 - y_2| \tag{7.21}$$

for all $-\infty < y_1, y_2 < \infty$, $t_0 \leq t \leq b$, and for some constant $K > 0$. Let the initial errors satisfy

$$\eta(h) \equiv \max_{0 \leq i \leq p} |Y(t_i) - y_h(t_i)| \to 0 \quad \text{as } h \to 0. \tag{7.22}$$

Assume that the solution $Y(t)$ is continuously differentiable and the method is consistent, that is, that it satisfies (7.5). Finally, assume that the coefficients a_j are all nonnegative,

$$a_j \geq 0, \qquad j = 0, 1, \ldots, p. \tag{7.23}$$

Then the method (7.1) is convergent and

$$\max_{t_0 \leq t_n \leq b} |Y(t_n) - y_h(t_n)| \leq c_1 \eta(h) + c_2 \tau(h) \tag{7.24}$$

for suitable constants c_1, c_2. If the solution $Y(t)$ is $m + 1$ times continuously differentiable, the method (7.1) is of order m, and the initial errors satisfy $\eta(h) = \mathcal{O}(h^m)$, then the order of convergence of the method is m; that is, the error is of size $\mathcal{O}(h^m)$.

Proof. Rewrite (7.3), and use $Y'(t) = f(t, Y(t))$ to get

$$Y(t_{n+1}) = \sum_{j=0}^{p} a_j Y(t_{n-j}) + h \sum_{j=-1}^{p} b_j f(t_{n-j}, Y(t_{n\ j})) + h\tau_n(Y).$$

Subtracting (7.1) from this equality and using the notation $e_i = Y(t_i) - y_i$, we obtain

$$e_{n+1} = \sum_{j=0}^{p} a_j e_{n-j} + h \sum_{j=-1}^{p} b_j \left[f(t_{n-j}, Y_{n-j}) - f(t_{n-j}, y_{n-j}) \right] + h\tau_n(Y).$$

Apply the Lipschitz condition (7.21) and the assumption (7.23) to obtain

$$|e_{n+1}| \leq \sum_{j=0}^{p} a_j |e_{n-j}| + hK \sum_{j=-1}^{p} |b_j| |e_{n-j}| + h\tau(h).$$

Introduce the following error bounding function

$$f_n = \max_{0 \leq i \leq n} |e_i|, \qquad n = 0, 1, \ldots, N(h).$$

Using this function, we have

$$|e_{n+1}| \leq \sum_{j=0}^{p} a_j f_n + hK \sum_{j=-1}^{p} |b_j| f_{n+1} + h\tau(h),$$

and applying (7.7), we obtain

$$|e_{n+1}| \leq f_n + hc f_{n+1} + h\tau(h), \qquad c = K \sum_{j=-1}^{p} |b_j|.$$

The right side is trivially a bound for f_n and thus

$$f_{n+1} \leq f_n + hc f_{n+1} + h\tau(h).$$

For $hc \leq \frac{1}{2}$, which is true as $h \to 0$, we obtain

$$f_{n+1} \leq \frac{f_n}{1 - hc} + \frac{h}{1 - hc}\tau(h)$$

$$\leq (1 + 2hc)f_n + 2h\tau(h).$$

Noting that $f_p = \eta(h)$, proceed as in the proof of Theorem 2.4 in Chapter 2, from (2.25) onward. Then

$$f_n \leq e^{2c(b-t_0)}\eta(h) + \left[\frac{e^{2c(b-t_0)} - 1}{c}\right]\tau(h), \qquad t_0 \leq t_n \leq b. \qquad (7.25)$$

This completes the proof.   ∎

To obtain a rate of convergence of $\mathcal{O}(h^m)$ for the method (7.1), it is necessary that each step have an error

$$T_n(Y) = \mathcal{O}(h^{m+1}).$$

But the initial values $y_0, \ldots, y_p$ need to be computed with an accuracy of only $\mathcal{O}(h^m)$, since $\eta(h) = \mathcal{O}(h^m)$ is sufficient in (7.24).

The result (7.25) can be improved somewhat for particular cases, but the order of convergence will remain the same. As with Euler's method, a complete stability analysis can be given, yielding a result of the form (2.49) in Chapter 2. The analysis is a straightforward modification of that described in Section 2.4 of Chapter 2. Similarly, an asymptotic error analysis can also be given.

7.3 A GENERAL ERROR ANALYSIS

We begin with a few definitions. The concept of *stability* was introduced with Euler's method, and we now generalize it. Let $\{y_n : 0 \leq n \leq N(h)\}$ denote the solution of (7.1) with initial values $y_0, y_1, \ldots, y_p$ for some differential equation $Y'(t) = f(t, Y(t))$ and for all sufficiently small values of h, say $h \leq h_0$. Recall that $N(h)$ denotes the largest subscript N for which $t_N \leq b$. For each $h \leq h_0$, perturb the initial values $y_0, \ldots, y_p$ to new values $z_0, \ldots, z_p$ with

$$\max_{0 \leq n \leq p} |y_n - z_n| \leq \epsilon. \qquad (7.26)$$

Note that these initial values are allowed to depend on h. We say that the family of discrete numerical solutions $\{y_n : 0 \leq n \leq N(h)\}$, obtained from (7.1), is *stable* if there is a constant c, independent of $h \leq h_0$ and valid for all sufficiently small ϵ, for which

$$\max_{0 \leq n \leq N(h)} |y_n - z_n| \leq c\epsilon, \qquad 0 < h \leq h_0. \qquad (7.27)$$

Consider all differential equation problems

$$Y'(t) = f(t, Y(t)), \quad t \geq t_0,$$
$$Y(t_0) = Y_0 \qquad\qquad\qquad\qquad (7.28)$$

with the derivative function $f(t, z)$ continuous and satisfying the Lipschitz condition (7.21). Suppose further that the approximating solutions $\{y_n\}$ are all stable. Then we say that (7.1) is a *stable numerical method*.

To define *convergence* for a given problem (7.28), suppose that the initial values $y_0, \ldots, y_p$ satisfy

$$\eta(h) \equiv \max_{0 \leq n \leq p} |Y(t_n) - y_n| \to 0 \quad \text{as } h \to 0. \tag{7.29}$$

Then the solution $\{y_n\}$ is said to converge to $Y(t)$ if

$$\max_{t_0 \leq t_n \leq b} |Y(t_n) - y_n| \to 0 \quad \text{as } h \to 0. \tag{7.30}$$

If (7.1) is convergent for all problems (7.28) with the properties specified immediately following (7.28), then it is called a *convergent numerical method*. Convergence can be shown to imply consistency; consequently, we consider only methods satisfying (7.7)–(7.8). The necessity of the condition (7.7) follows from the assumption of convergence of (7.1) for the problem

$$Y'(t) \equiv 0, \quad Y(0) = 1.$$

Just take $y_0 = \cdots = y_p = 1$, and observe the consequences of the convergence of y_{p+1} to $Y(t) \equiv 1$. We leave the proof of the necessity of (7.8) as Problem 8.

The convergence and stability of (7.1) are linked to the roots of the polynomial

$$\rho(r) = r^{p+1} - \sum_{j=0}^{p} a_j r^{p-j}. \tag{7.31}$$

Note that $\rho(1) = 0$ from the consistency condition (7.7). Let $r_0, \ldots, r_p$ denote the roots of $\rho(r)$, repeated according to their multiplicity, and let $r_0 = 1$. The method (7.1) satisfies the *root condition* if

$$(R1) \qquad |r_j| \leq 1, \qquad j = 0, 1, \ldots, p, \tag{7.32}$$

$$(R2) \qquad |r_j| = 1 \Longrightarrow \rho'(r_j) \neq 0. \tag{7.33}$$

The first condition requires all roots of $\rho(r)$ to lie on the unit circle $\{z \colon |z| \leq 1\}$ in the complex plane. Condition (7.33) states that all roots on the boundary of the circle are to be simple roots of $\rho(r)$.

7.3.1 Stability theory

All of the numerical methods presented in the preceding chapters have been stable, but we now give an example of a consistent unstable multistep method. This is to motivate the need to develop a general theory of stability.

Example 7.5 Consider the two step method

$$y_{n+1} = 3y_n - 2y_{n-1} + \tfrac{1}{2}h\left[f(t_n, y_n) - 3f(t_{n-1}, y_{n-1})\right], \quad n \geq 1. \tag{7.34}$$

It can be shown to have the truncation error

$$T_n(Y) = \tfrac{7}{12}h^3 Y^{(3)}(\xi_n), \qquad t_{n-1} \leq \xi_n \leq t_{n+1}$$

and therefore, it is a consistent method. Consider solving the problem $Y'(t) \equiv 0$, $Y(0) = 0$, which has the solution $Y(t) \equiv 0$. Using $y_0 = y_1 = 0$, the numerical solution is clearly $y_n = 0$, $n \geq 0$. Perturb the initial data to $z_0 = \epsilon/2$, $z_1 = \epsilon$, for some $\epsilon \neq 0$. Then the corresponding numerical solution can be shown to be

$$z_n = \epsilon \cdot 2^{n-1}, \qquad n \geq 0. \tag{7.35}$$

The reader should check this assertion. To see the effect of the perturbation on the original solution, let us assume that

$$\max_{t_0 \leq t_n \leq b} |y_n - z_n| = \max_{0 \leq t_n \leq b} |\epsilon| \, 2^{n-1} = |\epsilon| \, 2^{N(h)-1}.$$

Since $N(h) \to \infty$ as $h \to 0$, the deviation of $\{z_n\}$ from $\{y_n\}$ increases as $h \to 0$. The method (7.34) is unstable, and it should never be used. Also, note that the root condition is violated, since $\rho(r) = r^2 - 3r + 2$ has the roots $r_0 = 1$, $r_1 = 2$. ∎

To investigate the stability of (7.1), we consider only the special equation

$$\begin{aligned} Y'(t) &= \lambda Y(t), \quad t \geq 0, \\ Y(0) &= 1 \end{aligned} \tag{7.36}$$

with the solution $Y(t) = e^{\lambda t}$; λ is allowed to be complex. This is the model problem of (4.3), and its use was discussed in Chapter 4. The results obtained will transfer to the study of stability for a general differential equation problem. An intuitive reason for this is easily derived. Expand $Y'(t) = f(t, Y(t))$ about (t_0, Y_0) to obtain

$$Y'(t) \approx f(t_0, Y_0) + f_t(t_0, Y_0)(t - t_0) + f_y(t_0, Y_0)(Y(t) - Y_0)$$

$$= \lambda(Y(t) - Y_0) + g(t) \tag{7.37}$$

with $\lambda = f_y(t_0, Y_0)$ and $g(t) = f(t_0, Y_0) + f_t(t_0, Y_0)(t - t_0)$. This is a valid approximation if $|t - t_0|$ is sufficiently small. Introducing $V(t) = Y(t) - Y_0$,

$$V'(t) \approx \lambda V(t) + g(t). \tag{7.38}$$

The inhomogeneous term $g(t)$ will drop out of all derivations concerning numerical stability, because we are concerned with differences of solutions of the equation. Dropping $g(t)$ in (7.38), we obtain the model equation (7.36).

In the case that $\mathbf{Y}' = \mathbf{f}(t, \mathbf{Y})$ represents a system of m differential equations, which is discussed in Chapter 3, the partial derivative $\mathbf{f}_y(t, \mathbf{y})$ becomes a Jacobian matrix,

$$[\mathbf{f}_y(t, \mathbf{y})]_{ij} = \frac{\partial f_i}{\partial y_j}, \qquad 1 \le i, j \le m.$$

Thus the model equation becomes

$$\mathbf{y}' = \Lambda \mathbf{y} + \mathbf{g}(t), \tag{7.39}$$

a system of m linear differential equations with $\Lambda = \mathbf{f}_y(t_0, \mathbf{Y}_0)$. It can be shown that in many cases, this system reduces to an equivalent system

$$z_i' = \lambda_i z_i + \gamma_i(t), \qquad 1 \le i \le m \tag{7.40}$$

with $\lambda_1, \dots, \lambda_m$ the eigenvalues of Λ (see Problem 6). With (7.40), we are back to the simple model equation (7.36), provided we allow λ to be complex in order to include all possible eigenvalues of Λ.

Applying (7.1) to the model equation (7.36), we obtain

$$y_{n+1} = \sum_{j=0}^{p} a_j y_{n-j} + h\lambda \sum_{j=-1}^{p} b_j y_{n-j}, \tag{7.41}$$

$$(1 - h\lambda b_{-1}) y_{n+1} - \sum_{j=0}^{p} (a_j + h\lambda b_j) y_{n-j} = 0, \qquad n \ge p. \tag{7.42}$$

This is a *homogeneous linear difference equation* of order $p + 1$, and the theory for its solvability is completely analogous to that of $(p + 1)$-order homogeneous linear differential equations. As a general reference, see Henrici [45, pp. 210–215] or Isaacson and Keller [47, pp. 405–417].

We attempt to find a general solution by first looking for solutions of the special form

$$y_n = r^n, \qquad n \ge 0.$$

If we can find $p+1$ linearly independent solutions, then an arbitrary linear combination will give the general solution of (7.42).

Substituting $y_n = r^n$ into (7.42) and canceling r^{n-p}, we obtain

$$(1 - h\lambda b_{-1}) r^{p+1} - \sum_{j=0}^{p} (a_j + h\lambda b_j) r^{p-j} = 0. \tag{7.43}$$

This is called the *characteristic equation,* and the left-side is the *characteristic polynomial.* The roots are called *characteristic roots.* Define

$$\sigma(r) = b_{-1} r^{p+1} + \sum_{j=0}^{p} b_j r^{p-j},$$

and recall the definition (7.31) of $\rho(r)$. Then (7.43) becomes

$$\rho(r) - h\lambda\sigma(r) = 0. \tag{7.44}$$

Denote the characteristic roots by

$$r_0(h\lambda), \ldots, r_p(h\lambda),$$

which can be shown to depend continuously on the value of $h\lambda$. When $h\lambda = 0$, equation (7.44) becomes simply $\rho(r) = 0$, and we have $r_j(0) = r_j$, $j = 0, 1, \ldots, p$ for the earlier roots r_j of $\rho(r) = 0$. Since $r_0 = 1$ is a root of $\rho(r)$, we let $r_0(h\lambda)$ be the root of (7.44) for which $r_0(0) = 1$. The root $r_0(h\lambda)$ is called the *principal root* for reasons that will become apparent later. If the roots $r_j(h\lambda)$ are all distinct, then the general solution of (7.42) is

$$y_n = \sum_{j=0}^{p} \gamma_j \left[r_j(h\lambda)\right]^n, \qquad n \geq 0. \tag{7.45}$$

But if

$$r_j(h\lambda) = r_{j+1}(h\lambda) = \cdots = r_{j+\nu-1}(h\lambda)$$

is a root of multiplicity $\nu > 1$, then the following are ν linearly independent solutions of (7.42):

$$\{[r_j(h\lambda)]^n\}, \ \{n\,[r_j(h\lambda)]^n\}, \ \ldots, \ \{n^{\nu-1}\,[r_j(h\lambda)]^n\}.$$

Moreover, in the formula (7.45), the part

$$\gamma_j\,[r_j(h\lambda)]^n + \cdots + \gamma_{j+\nu-1}\,[r_{j+\nu-1}(h\lambda)]^n$$

needs to be replaced by

$$[r_j(h\lambda)]^n \left(\gamma_j + \gamma_{j+1}n + \cdots + \gamma_{j+\nu-1}n^{\nu-1}\right). \tag{7.46}$$

These can be used with the solution arising from the other roots to generate a general solution for (7.42), comparable to (7.45).

In particular, for consistent methods it can be shown that

$$[r_0(h\lambda)]^n = e^{\lambda t_n} + \mathcal{O}(h) \tag{7.47}$$

as $h \to 0$. The remaining roots $r_1(h\lambda), \ldots, r_p(h\lambda)$ are called *parasitic roots* of the numerical method. The term

$$\sum_{j=1}^{p} \gamma_j\,[r_j(h\lambda)]^n \tag{7.48}$$

is called a *parasitic solution*. It is a creation of the numerical method and does not correspond to any solution of the original differential equation being solved.

Theorem 7.6 *Assume the consistency conditions (7.7)–(7.8). Then the multistep method (7.1) is stable if and only if the root condition (7.32)–(7.33) is satisfied.*

The proof makes essential use of the general solution (7.45) in the case of distinct roots $\{r_j(h\lambda)\}$, or the variant of (7.45) modified according to (7.46) when multiple roots are present. The reader is referred to [11, p. 398] for a partial proof and to [47, pp. 405-417] for a more complete development.

7.3.2 Convergence theory

The following result generalizes Theorem 7.4 from earlier in this chapter, giving necessary and sufficient conditions for the convergence of multistep methods.

Theorem 7.7 *Assume the consistency conditions (7.7)–(7.8). Then the multistep method (7.1) is convergent if and only if the root condition (7.32)–(7.33) is satisfied.*

Again, we refer the reader to [11, p. 401] for a partial proof and to [47, pp. 405–417] for a more complete development.

The following is a well-known result, and it is a trivial consequence of Theorems 7.6 and 7.7.

Corollary 7.8 *Let (7.1) be a consistent multistep method. Then it is convergent if and only if it is stable.*

Example 7.9 Return to the two-step methods of order 2, developed in Example 7.3. The polynomial $\rho(r)$ is given by

$$\rho(r) = r^2 - a_0 r - a_1, \qquad a_0 + a_1 = 1.$$

Then

$$\rho(r) = (r - 1)(r + 1 - a_0),$$

and the roots are

$$r_0 = 1, \qquad r_1 = a_0 - 1.$$

The root condition requires

$$-1 \le a_0 - 1 < 1,$$
$$0 \le a_0 < 2,$$

to ensure convergence and stability of the associated two step method in (7.16). ∎

7.3.3 Relative stability and weak stability

Consider again the model equation (7.36) and its numerical solution (7.45). For a convergent numerical method, it can be shown that in the general solution (7.45), we obtain

$$\gamma_0 \to 1,$$
$$\gamma_j \to 0, \quad j = 1, \dots, p$$

as $h \to 0$. The parasitic solution (7.48) converges to zero as $h \to 0$, and the term $\gamma_0 \left[r_0(h\lambda) \right]^n$ converges to $Y(t) = e^{\lambda t}$ with $t_n = t$ fixed. However, for a fixed h with increasing t_n, we also would like the parasitic solution to remain small relative to the principal part of the solution $\gamma_0[r_0(h\lambda)]^n$. This will be true if the characteristic roots satisfy

$$|r_j(h\lambda)| \le r_0(h\lambda), \qquad j = 1, 2, \dots, p \tag{7.49}$$

for all sufficiently small values of h. This leads us to the definition of relative stability.

We say that the method (7.1) is *relatively stable* if the characteristic roots $r_j(h\lambda)$ satisfy (7.49) for all sufficiently small nonzero values of $|h\lambda|$. Further, the method is said to satisfy the *strong root condition* if

$$|r_j(0)| < 1, \qquad j = 1, 2, \ldots, p. \tag{7.50}$$

This condition is easy to check, and it implies relative stability. Just use the continuity of the roots $r_j(h\lambda)$ with respect to $h\lambda$ to verify that (7.50) implies (7.49). Relative stability does not imply the strong root condition, although they are equivalent for most methods. If a multistep method is stable but not relatively stable, then it will be called *weakly stable*.

Example 7.10

(1) For the midpoint method, we obtain

$$r_0(h\lambda) = 1 + h\lambda + \mathcal{O}(h^2), \qquad r_1(h\lambda) = -1 + h\lambda + \mathcal{O}(h^2). \tag{7.51}$$

For $\lambda < 0$, we have
$$|r_1(h\lambda)| > r_0(h\lambda)$$

for all small values of $h > 0$, and thus (7.49) is not satisfied. The midpoint method is not relatively stable; it is only weakly stable. We leave it as an exercise to show experimentally that the midpoint method has undesirable stability when $\lambda < 0$ for the model equation (7.28).

(2) The Adams–Bashforth and Adams–Moulton methods of Chapter 6 have the same characteristic polynomial when $h = 0$,

$$\rho(r) = r^{p+1} - r^p. \tag{7.52}$$

The roots are $r_0 = 1, r_j = 0, j = 1, 2, \ldots, p$; thus the strong root condition is satisfied and the Adams methods are relatively stable. ∎

PROBLEMS

1. Consider the two-step method

$$y_{n+1} = \frac{1}{2}(y_n + y_{n-1}) + \frac{h}{4}\left[4y'_{n+1} - y'_n + 3y'_{n-1}\right], \qquad n \geq 1$$

with $y'_n \equiv f(t_n, y_n)$. Show that it has order 2, and find the leading term in the truncation error, written as in (7.15).

2. Recall the midpoint method

$$y_{n+1} = y_{n-1} + 2hf(t_n, y_n), \qquad n \geq 1$$

from Problem 11 in Chapter 6.

(a) Show that the midpoint method has order 2, as noted earlier following (7.2).

(b) Show that the midpoint method is not relatively stable.

3. Write a program to solve $Y'(t) = f(t, Y(t))$, $Y(t_0) = Y_0$ using the midpoint rule of Problem 2. Use a fixed stepsize h. For the initial value y_1, use the Euler method with $y_0 = Y_0$,

$$y_1 = y_0 + hf(t_0, y_0).$$

Using the program, solve the problem

$$Y'(t) = -Y(t) + 2\cos(t), \qquad Y(0) = 1.$$

The true solution is $Y(t) = \cos(t) + \sin(t)$. Solve this problem on the interval $[0, 10]$, and use stepsizes of $h = 0.2, 0.1, 0.05$. Comment on your results. Produce a graph of the error.

4. Show that the two-step method

$$y_{n+1} = -y_n + 2y_{n-1} + h\left[\tfrac{5}{2}y'_n + \tfrac{1}{2}y'_{n-1}\right], \qquad n \geq 1$$

is of order 2 and unstable. Also, show directly that it need not converge when solving $Y'(t) = f(t, Y(t))$ by considering the special problem

$$Y'(t) = 0, \qquad Y(0) = 0.$$

For the numerical method, consider using the initial values

$$y_0 = h, \quad y_1 = -2h.$$

Hint: Use the general formula (7.45), and examine the numerical solution for $t_n = nh = 1$.

5. Consider the general formula for all explicit two-step methods,

$$y_{n+1} = a_0 y_n + a_1 y_{n-1} + h\left[b_0 f(t_n, y_n) + b_1 f(t_{n-1}, y_{n-1})\right], \qquad n \geq 1.$$

(a) Consider finding all such two-step methods that are of order 2. Show that the coefficients must satisfy the equations

$$a_0 + a_1 = 1, \qquad -a_1 + b_0 + b_1 = 1, \qquad a_1 - 2b_1 = 1.$$

Solve for $\{a_1, b_0, b_1\}$ in terms of a_0.

(b) Find a formula for the leading term in the truncation error, written as in (7.15). It will depend on a_0.

(c) What are the restrictions on a_0 for this two-step method to be stable? To be convergent?

6. Consider the model equation (7.39) with Λ, a square matrix of order m. Assume $\Lambda = P^{-1}DP$ with D a diagonal matrix with entries $\lambda_1, \ldots, \lambda_m$. Introduce the new unknown vector function $\mathbf{z} = P\mathbf{y}(t)$. Show that (7.39) converts to the form given in (7.40), demonstrating the reduction to the one-dimensional model equation.

   Hint: In (7.39) replace Λ with $P^{-1}DP$, and then introduce the new unknowns $\mathbf{z} = P\mathbf{y}$. Simplify to a differential equation for $\mathbf{z}$.

7. For solving $Y'(t) = f(t, Y(t))$, consider the numerical method

$$y_{n+1} = y_n + \frac{h}{2}\left[y_n' + y_{n+1}'\right] + \frac{h^2}{12}\left[y_n'' - y_{n+1}''\right], \qquad n \geq 0.$$

   Here $y_n' = f(t_n, y_n)$,

$$y_n'' = \frac{\partial f(t_n, y_n)}{\partial t} + f(t_n, y_n)\left.\frac{\partial f(t_n, y)}{\partial y}\right|_{z-y_n}$$

   with this formula based on differentiating $Y'(t) = f(t, Y(t))$.

   (a) Show that this is a fourth-order method with $T_n(Y) = \mathcal{O}(h^5)$.
       Hint: Use the Taylor approximation method used earlier in deriving the results of Theorem 7.2, modifying this procedure as necessary for analyzing this case.

   (b) Show that the region of absolute stability contains the entire negative real axis of the complex $h\lambda$-plane.

8. Prove that (7.8) is necessary for the multistep numerical method (7.1) to be consistent.
   Hint: Apply (7.1) to the initial value problem

$$Y'(t) = 1, \qquad Y(0) = 0$$

   with exact initial conditions.

9. (a) Find all explicit fourth-order formulas of the form

$$y_{n+1} = a_0 y_n + a_1 y_{n-1} + a_2 y_{n-2}$$
$$+ h\left[b_0 y_n' + b_1 y_{n-1}' + b_2 y_{n-2}'\right], \qquad n \geq 2.$$

   (b) Show that every such method is unstable.

10. (a) Consider methods of the form

$$y_{n+1} = y_{n-q} + h\sum_{j=-1}^{p} b_j f(x_{n-j}, y_{n-j})$$

with $q \geq 1$. Show that such methods do not satisfy the strong root condition. As a consequence, most such methods are only weakly stable.

(b) Find an example with $q = 1$ that is relatively stable.

11. For the polynomial $\rho(r) = r^{p+1} - \sum_{j=0}^{p} a_j r^{p-j}$, assume $a_j \geq 0, 0 \leq j \leq p$, and $\sum_{j=0}^{p} a_j = 1$. Show that the roots of $\rho(r)$ will satisfy the root conditions (7.32) and (7.33). This shows directly that Theorem 7.4 is a corollary of Theorem 7.7.

CHAPTER 8

STIFF DIFFERENTIAL EQUATIONS

The numerical solution of stiff differential equations is a widely studied subject. Such equations (including systems of differential equations) appear in a wide variety of applications, in subjects as diverse as chemical kinetics, mechanical systems, and the numerical solution of partial differential equations. In this section, we sketch some of the main ideas about this subject, and we show its relation to the numerical solution of the simple heat equation from partial differential equations.

There are several definitions of the concept of stiff differential equation. The most important common feature of these definitions is that when such equations are being solved with standard numerical methods (e.g., the Adams–Bashforth methods of Chapter 6), the stepsize h must be extremely small in order to maintain stability — far smaller than would appear to be necessary from a consideration of the truncation error. A numerical illustration for Euler's method is given in Table 4.3 as a part of Example 4.2 in Chapter 4.

Definitions and results related to the topic of stiff differential equations were introduced in Chapter 4 (see (4.3)–(4.5) and (4.10)) and Chapter 6 (see the discussion accompanying (6.19)–(6.20)). For convenience, we review those ideas here. As was discussed preceding (4.3) in Chapter 4, the following model problem is used to test the performance of numerical methods,

$$Y' = \lambda Y, \quad t > 0,$$
$$Y(0) = 1. \tag{8.1}$$

Following (7.36) in Chapter 7, a derivation was given to show that (8.1) is useful in studying the stability of numerical methods for very general systems of nonlinear differential equations; we review this in more detail in a later paragraph.

When the constant λ is real, we assume $\lambda < 0$; or more generally, when λ is complex, we assume $\text{Real}(\lambda) < 0$. This assumption about λ is generally associated with stable differential equation problems (see Section 1.2). The true solution of the model problem is

$$Y(t) = e^{\lambda t}. \tag{8.2}$$

From our assumption on λ, we have

$$Y(t) \to 0 \quad \text{as} \quad t \to \infty. \tag{8.3}$$

The kind of stability property we would prefer for a numerical method is that when it is applied to (8.1), the numerical solution satisfies

$$y_h(t_n) \to 0 \quad \text{as} \quad t_n \to \infty \tag{8.4}$$

for any choice of the stepsize h. Such numerical methods are called *absolutely stable* or *A-stable*. For an arbitrary numerical method, the set of values $h\lambda$ for which (8.4) is satisfied, considered as a subset of the complex plane, is called the *region of absolute stability* of the numerical method. The dependence on the product $h\lambda$ is based on the general solution to the finite difference method for solving (8.1), given in (7.45) of Chapter 7.

Example 8.1 We list here the region of absolute stability as derived in earlier chapters. Again, we consider only λ satisfying our earlier assumption that $\text{Real}(\lambda) < 0$.

- For Euler's method, it was shown following (4.5) that (8.4) is satisfied if and only if

$$|1 + h\lambda| = |h\lambda - (-1)| < 1. \tag{8.5}$$

Thus $h\lambda$ is in the region of absolute stability if and only if it is within a distance of 1 from the point -1 in the complex plane. The region of absolute stability is a circle of unit radius with center at -1. For real λ, this requires

$$-2 < h\lambda < 0.$$

- For the backward Euler method of (4.9), it was shown in and following (4.10) that (8.4) is satisfied for every value of $h\lambda$ in which $\text{Real}(\lambda) < 0$. The backward Euler method is A-stable.

- For the trapezoidal method of (4.22), it was left to Problem 2 in Chapter 4 to show that (8.4) is satisfied for every value of $h\lambda$ in which $\text{Real}(\lambda) < 0$. The trapezoidal method is A-stable.

- For the Adams–Bashforth method of order 2, namely

$$y_{n+1} = y_n + \frac{h}{2}[3y_n' - y_{n-1}'], \qquad n \geq 1 \tag{8.6}$$

(see Table 6.2), it was stated in Example 6.6 that the real part of the region of absolute stability is the interval

$$-1 < h\lambda < 0. \tag{8.7}$$

∎

Why is this of interest? If a method is absolutely stable, then there are no restrictions on h in order for the numerical method to be stable in the sense of (8.4). However, consider what happens to the stepsize h if a method has a region of absolute stability that is bounded (and say, of moderate size). Suppose that the value of λ has a real part that is negative and of very large magnitude. Then h must be correspondingly small for $h\lambda$ to belong to the region of absolute stability of the method. Even if the truncation error is small, it is necessary that $h\lambda$ belong to the region of absolute stability to ensure that the error in the approximate solution $\{y_n\}$ is also small.

Example 8.2 Recall Example 4.2 in Chapter 4, which illustrated the computational effects of regions of absolute stability for the Euler, backward Euler, and trapezoidal methods when solving the problem

$$Y'(t) = \lambda Y(t) + (1 - \lambda)\cos(t) - (1 + \lambda)\sin(t), \qquad Y(0) = 1. \tag{8.8}$$

The true solution is $Y(t) = \sin(t) + \cos(t)$. We augment those earlier calculations by giving results for the Adams–Bashforth method (8.6) when solving (8.8). For simplicity, we use $y_1 = Y(t_1)$. Numerical results for several values of λ are given in Table 8.1. The values of h are the same as those used in Table 4.3 for Euler's method in Example 4.2. The stability of the error in the numerical results are consistent with the region of absolute stability given in (8.7). ∎

Returning to the derivation following (7.36) in Chapter 7, we looked at the linearization of the system

$$\mathbf{Y}' = \mathbf{f}(t, \mathbf{Y}) \tag{8.9}$$

of m differential equations, resulting in the approximating linear system

$$\mathbf{Y}' = \Lambda \mathbf{Y} + \mathbf{g}(t). \tag{8.10}$$

In this, $\Lambda = \mathbf{f}_y(t_0, \mathbf{Y}_0)$ is the $m \times m$ Jacobian matrix of $\mathbf{f}$ evaluated at $(t_0, \mathbf{Y}_0)$. As was explored in Problem 6 of Chapter 7, many such systems can be reduced to a set of m independent scalar equations

$$Y_i' = \lambda_i Y_i + g_i(t), \qquad i = 1, \ldots, m.$$

Table 8.1   The Adams-Bashforth method (8.6) for solving (8.8)

λ	t	Error $h = 0.5$	Error $h = 0.1$	Error $h = 0.01$
-1	1	$-2.39e - 2$	$-7.58e - 4$	$-7.24e - 6$
	2	$4.02e - 2$	$2.13e - 3$	$2.28e - 5$
	3	$1.02e - 1$	$4.31e - 3$	$4.33e - 5$
	4	$8.50e - 2$	$2.98e - 3$	$2.82e - 5$
	5	$-3.50e - 3$	$-9.16e - 4$	$-1.13e - 5$
-10	1	$-2.39e - 2$	$-1.00e - 4$	$6.38e - 7$
	2	$-1.10e + 0$	$3.75e - 4$	$5.25e - 6$
	3	$-5.23e + 1$	$3.83e - 4$	$5.03e - 6$
	4	$2.46e + 3$	$-8.32e - 5$	$1.91e - 7$
	5	$-1.16e + 5$	$-5.96e - 4$	$-4.83e - 6$
-50	1	$-2.39e - 2$	$-1.57e + 3$	$2.21e - 7$
	2	$3.25e + 1$	$-3.64e + 11$	$1.09e - 6$
	3	$4.41e + 4$	$-8.44e + 19$	$9.60e - 7$
	4	$-5.98e + 7$	$-1.96e + 28$	$-5.54e - 8$
	5	$-8.12e + 10$	$-4.55e + 36$	$-1.02e - 6$

As was discussed following (7.36), this leads us back to the model equation (8.1) with λ an eigenvalue of the Jacobian matrix $f_y(t_0, Y_0)$.

We say that the differential equation $Y' = f(t, Y)$ is *stiff* if some of the eigenvalues λ_j of Λ, or more generally of $f_y(t, Y)$, have a negative real part of very large magnitude. The question may arise as to how large the eigenvalue should be to be considered large? The magnitude of the eigenvalues might depend on the units of measurement used, for example, which has no impact on the amount of computation needed to accurately solve a particular problem. The crucial test is to consider the eigenvalue(s) associated with the slowest rates of change, and compare them with the eigenvalue(s) associated with the fastest rates of change. A simple test is to look at the ratio $\max_i |\lambda_i| / \min_i |\lambda_i|$. If this number is large, then the problem is stiff. For example, in the pendulum model (3.13), the two eigenvalues in the linearization have the same or similar magnitudes. So it is not a stiff problem. Most problems that we have seen so far are not stiff. Yet, stiff problems are common in practice. In the next section we see one very important example.

We study numerical methods for stiff equations by considering their effect on the model equation (8.1). This approach has its limitations, some of which we indicate later, but it does give us a means of rejecting unsatisfactory methods, and it suggests some possibly satisfactory methods. Before giving some higher-order methods that are suitable for solving stiff differential equations, we give an important practical example.

8.1 THE METHOD OF LINES FOR A PARABOLIC EQUATION

Consider the following parabolic partial differential equation problem:

$$U_t = U_{xx} + G(x,t), \qquad 0 < x < 1, \qquad t > 0, \tag{8.11}$$

$$U(0,t) = d_0(t), \qquad U(1,t) = d_1(t), \qquad t \geq 0, \tag{8.12}$$

$$U(x,0) = f(x), \qquad 0 \leq x \leq 1. \tag{8.13}$$

The unknown function $U(x,t)$ depends on the time t and a spatial variable x, and $U_t = \partial U / \partial t$, $U_{xx} = \partial^2 U / \partial x^2$. The conditions (8.12) are called *boundary conditions*, and (8.13) is called an *initial condition*. The solution U can be interpreted as the temperature of an insulated rod of length 1 with $U(x,t)$, the temperature at position x and time t; thus (8.11) is often called the *heat equation*. The functions G, d_0, d_1, and f are assumed given and smooth. For a development of the theory of (8.11)–(8.13), see Widder [78] or any standard introduction to partial differential equations. We give the *method of lines* for solving for U, a popular numerical method for solving numerically linear and nonlinear partial differential equations of parabolic type. This numerical method also leads to the necessity of solving a stiff system of ordinary differential equations.

Let $m > 0$ be an integer, define $\delta = 1/m$, and define the spatial nodes

$$x_j = j\delta, \qquad j = 0, 1, \dots, m.$$

We discretize (8.11) by approximating the spatial derivative U_{xx} in the equation. Using a standard result in the theory of numerical differentiation,

$$U_{xx}(x_j, t) = \frac{U(x_{j+1}, t) - 2U(x_j, t) + U(x_{j-1}, t)}{\delta^2} - \frac{\delta^2}{12} \frac{\partial^4 U(\xi_j, t)}{\partial x^4} \tag{8.14}$$

for $j = 1, 2, \dots, m - 1$, where each $\xi_j \equiv \xi_j(t)$ is some point between x_{j-1} and x_{j+1}. For a derivation of this formula, see [11, p. 318] or [12, p. 237]. Substituting into (8.11), we obtain

$$U_t(x_j, t) = \frac{U(x_{j+1}, t) - 2U(x_j, t) + U(x_{j-1}, t)}{\delta^2} + G(x_j, t)$$
$$- \frac{\delta^2}{12} \frac{\partial^4 U(\xi_j, t)}{\partial x^4}, \qquad 1 \leq j \leq m - 1. \tag{8.15}$$

Equation (8.11) is to be approximated at each interior node point x_j.

We drop the final term in (8.15), the truncation error in the numerical differentiation. Forcing equality in the resulting approximate equation, we obtain

$$u_j'(t) = \frac{1}{\delta^2} [u_{j+1}(t) - 2u_j(t) + u_{j-1}(t)] + G(x_j, t) \tag{8.16}$$

for $j = 1, 2, \dots, m - 1$. The functions $u_j(t)$ are intended as approximations of $U(x_j, t)$, $1 \leq j \leq m - 1$. This is the *method of lines* approximation to (8.11), and it is a system of $m - 1$ ordinary differential equations. Note that $u_0(t)$ and $u_m(t)$, which are needed in (8.16) for $j = 1$ and $j = m - 1$, are given using (8.12):

$$u_0(t) = d_0(t), \qquad u_m(t) = d_1(t). \tag{8.17}$$

The initial condition for (8.16) is given by (8.13):

$$u_j(0) = f(x_j), \qquad 1 \le j \le m - 1. \tag{8.18}$$

The term *method of lines* comes from solving for $U(x, t)$ along the lines $(x_j, t), t \ge 0$, $1 \le j \le m - 1$ in the (x, t) plane.

Under suitable assumptions on the functions d_0, d_1, G, and f, it can be shown that

$$\max_{\substack{0 \le j \le m \\ 0 \le t \le T}} |U(x_j, t) - u_j(t)| \le C_T \delta^2. \tag{8.19}$$

Thus to complete the solution process, we need only solve the system (8.16).

It is convenient to write (8.16) in matrix form. Introduce

$$\mathbf{u}(t) = [u_1(t), \dots, u_{m-1}(t)]^T, \qquad \mathbf{u}_0 = [f(x_1), \dots, f(x_{m-1})]^T,$$

$$\mathbf{g}(t) = \left[\frac{d_0(t)}{\delta^2} + G(x_1, t), \ G(x_2, t), \ \dots, \ G(x_{m-2}, t), \ \frac{d_1(t)}{\delta^2} + G(x_{m-1}, t) \right]^T,$$

$$\Lambda = \frac{1}{\delta^2} \begin{bmatrix} -2 & 1 & 0 & \cdots & & 0 \\ 1 & -2 & 1 & & & 0 \\ & & \ddots & & & \vdots \\ \vdots & & 1 & -2 & 1 \\ 0 & \cdots & 0 & 1 & -2 \end{bmatrix}.$$

The matrix Λ is of order $m - 1$. In the definitions of $\mathbf{u}$ and $\mathbf{g}$, the superscript T indicates matrix transpose, so that $\mathbf{u}$ and $\mathbf{g}$ are column vectors of length $m - 1$. Using these matrices, equations (8.16)–(8.18) can be rewritten as

$$\mathbf{u}'(t) = \Lambda \mathbf{u}(t) + \mathbf{g}(t), \qquad \mathbf{u}(0) = \mathbf{u}_0. \tag{8.20}$$

If Euler's method is applied, we have the numerical method

$$\mathbf{V}_{n+1} = \mathbf{V}_n + h[\Lambda \mathbf{V}_n + \mathbf{g}(t_n)], \qquad \mathbf{V}_0 = \mathbf{u}_0 \tag{8.21}$$

with $t_n = nh$ and $\mathbf{V}_n \approx \mathbf{u}(t_n)$. This is a well-known numerical method for the heat equation, called the *simple explicit method*. We analyze the stability of (8.21) and some other methods for solving (8.20).

Equation (8.20) is in the form of the model equation, (8.10), and therefore we need the eigenvalues of Λ to examine the stiffness of the system. These eigenvalues are all real and are given by

$$\lambda_j = -\frac{4}{\delta^2} \sin^2\left(\frac{j\pi}{2m}\right), \qquad 1 \le j \le m - 1. \tag{8.22}$$

A proof (which we omit here) can be obtained by showing a relationship between the characteristic polynomial for Λ and Chebyshev polynomials. Directly examining (8.22), we have

$$\lambda_{m-1} \leq \lambda_j \leq \lambda_1, \tag{8.23}$$

with

$$\lambda_{m-1} = \frac{-4}{\delta^2} \sin^2\left(\frac{(m-1)\pi}{2m}\right) \approx \frac{-4}{\delta^2},$$

$$\lambda_1 = \frac{-4}{\delta^2} \sin^2\left(\frac{\pi}{2m}\right) \approx -\pi^2$$

with the approximations valid for larger m. As $\lambda_{m-1}/\lambda_1 \approx 4/(\pi\delta)^2$, it can be seen that (8.20) is a stiff system if δ is small.

Applying (8.23) and (8.5) to the analysis of stability in (8.21), we must have

$$|1 + h\lambda_j| < 1, \qquad j = 1, \ldots, m - 1.$$

Using (8.22), this leads to the equivalent statement

$$0 < \frac{4h}{\delta^2} \sin^2\left(\frac{j\pi}{2m}\right) < 2, \qquad 1 \leq j \leq m - 1.$$

This will be satisfied if $4h/\delta^2 \leq 2$ or

$$h \leq \tfrac{1}{2}\delta^2. \tag{8.24}$$

If δ is at all small, say $\delta = 0.01$, then the timestep h must be quite small to ensure stability.

In contrast to the restriction (8.24) with Euler's method, the backward Euler method has no such restriction since it is A-stable. Applying the backward Euler method, our approximation to (8.20) is

$$\mathbf{V}_{n+1} = \mathbf{V}_n + h\left[\Lambda\mathbf{V}_{n+1} + \mathbf{g}(t_{n+1})\right], \qquad \mathbf{V}_0 = \mathbf{u}_0. \tag{8.25}$$

This is called the *simple implicit method* for solving the heat equation. To solve this linear problem for $\mathbf{V}_{n+1}$, we rewrite the equation as

$$(I - h\Lambda)\mathbf{V}_{n+1} = \mathbf{V}_n + h\mathbf{g}(t_{n+1}). \tag{8.26}$$

Solving for $\mathbf{V}_{n+1}$ gives

$$\mathbf{V}_{n+1} = (I - h\Lambda)^{-1}\left[\mathbf{V}_n + h\mathbf{g}(t_{n+1})\right]. \tag{8.27}$$

Since all the eigenvalues λ_i of Λ are negative, the eigenvalues of $(I - h\Lambda)^{-1}$ are $1/(1 - h\lambda_i)$, which are all bounded by one. Because of this, the implicit Euler method for this problem is always stable; there is no limitation on the stepsize h, unlike the case for the explicit Euler method. Also, the linear system to be solved

Table 8.2 The method of lines: Euler's method ($h = \frac{1}{2}\delta^2$)

t	Error $m = 4$	Ratio	Error $m = 8$	Ratio	Error $m = 16$
1.0	$4.85e - 2$	4.096	$1.18e - 2$	4.024	$2.94e - 3$
2.0	$4.39e - 2$	4.096	$1.07e - 2$	4.024	$2.66e - 3$
3.0	$3.97e - 2$	4.096	$9.69e - 3$	4.024	$2.41e - 3$
4.0	$3.59e - 2$	4.096	$8.77e - 3$	4.024	$2.18e - 3$
5.0	$3.25e - 2$	4.096	$7.93e - 3$	4.024	$1.97e - 3$

Table 8.3 The method of lines: Backward Euler method ($h = 0.1$)

t	Error $m = 4$	Error $m = 8$	Error $m = 16$
1.0	$4.85e - 2$	$1.19e - 2$	$2.99e - 3$
2.0	$4.39e - 2$	$1.08e - 2$	$2.70e - 3$
3.0	$3.98e - 2$	$9.73e - 3$	$2.45e - 3$
4.0	$3.60e - 2$	$8.81e - 3$	$2.21e - 3$
5.0	$3.25e - 2$	$7.97e - 3$	$2.00e - 3$

is a tridiagonal system, and there is a well-developed numerical analysis for such linear systems (e.g. see [11, p. 527] or [12, p. 287]). It can be solved very rapidly with approximately $5m$ arithmetic operations per timestep, excluding the cost of computing the right side in (8.26). The cost of solving the Euler method (8.21) is almost as large, and thus the solution of (8.26) is not especially time-consuming.

Example 8.3 Solve the partial differential equation problem (8.11)–(8.13) with the functions G, d_0, d_1, and f, determined from the known solution

$$U = e^{-.1t} \sin(\pi x), \qquad 0 \leq x \leq 1, \quad t \geq 0. \tag{8.28}$$

Results for Euler's method (8.21) are given in Table 8.2, and results for the backward Euler method (8.25) are given in Table 8.3.

For Euler's method, we take $m = 4, 8, 16$, and to maintain stability, we take $h = \frac{1}{2}\delta^2$ from (8.24). This leads to the respective timesteps of $h \doteq 0.031, 0.0078, 0.0020$. From (8.19) and the error formula for Euler's method, we would expect the error to be proportional to δ^2, since $h = \frac{1}{2}\delta^2$. This implies that the error should decrease by a factor of 4 when m is doubled, and the results in Table 8.2 agree. In the table, the column "Error" denotes the maximum error at the node points (x_j, t), $0 \leq j \leq n$, for the given value of t.

For the solution of (8.20) by the backward Euler method, there need no longer be any connection between the spatial stepsize δ and the timestep h. By observing the error formula (8.19) for the method of lines and the truncation error formula (8.33) (use $p = 1$) for the backward Euler method, we see that the error in solving the problem (8.11)–(8.13) will be proportional to $h + \delta^2$. For the unknown function U of (8.26), there is a slow variation with t. Thus, for the truncation error associated with the time integration, we should be able to use a relatively large timestep h as compared to the spatial stepsize δ, for the two sources of error be relatively equal in size. In Table 8.3, we use $h = 0.1$ and $m = 4, 8, 16$. Note that this timestep is much larger than that used in Table 8.2 for Euler's method, and thus the backward Euler method is much more efficient for this example. ∎

For more discussion of the method of lines, see Aiken [1, pp. 124–148] and Schiesser [71].

8.1.1 MATLAB® programs for the method of lines

We give MATLAB programs for both the Euler method (8.21) and the backward Euler method (8.27).

Euler method code:

```
function [x,t,u] = MOL_Euler(d0,d1,f,G,T,h,m)
%
% function [x,t,u] = MOL_Euler(d0,d1,f,G,T,h,m)
%
% Use the method of lines to solve
%        u_t = u_xx + G(x,t), 0 < x < 1, 0 < t < T
% with boundary conditions
%        u(0,t) = d0(t), u(1,t) = d1(t)
% and initial condition
%        u(x,0) = f(x).
% Use Euler's method to solve the system of ODEs.
% For the discretization, use a spatial stepsize of
% delta=1/m and a timestep of h.
%
% For numerical stability, use a timestep of
% h = 1/(2*m^2) or smaller.

x = linspace(0,1,m+1)'; delta = 1/m; delta_sqr = delta^2;
t = (0:h:T)'; N = length(t);

% Initialize u.
u = zeros(m+1,N);
u(:,1) = f(x);
u(1,:)   = d0(t); u(m+1,:)   = d1(t);
```

```
% Solve for u using Euler's method.
for n=1:N-1
    g = G(x(2:m),t(n));
    u(2:m,n+1) = u(2:m,n) + (h/delta_sqr)*(u(1:(m-1),n) ...
                   - 2*u(2:m,n) + u(3:(m+1),n)) + h*g;
end
u = u';
end % MOL_Euler
```

Test of Euler method code:

```
function [x,t,u,error] = Test_MOL_Euler(index_u,t_max,h,m)

% Try this test program with
%     [x,t,u,error] = Test_MOL_Euler(2,5,1/128,8);

[x,t,u] = MOL_Euler(@d0,@d1,@f,@G,t_max,h,m);

% Graph numerical solution
[X,T] = meshgrid(x,t);
figure; mesh(X,T,u); shading interp
xlabel('x'); ylabel('t');
title(['Numerical solution u:  index of u = ',...
       num2str(index_u)])
disp('Press any key to continue.'); pause

% Graph error in numerical solution
true_u = true_soln(X,T); error = true_u - u;
disp(['Maximum error = ',num2str(max(max(abs(error))))])
figure; mesh(X,T,error); shading interp
xlabel('x'); ylabel('t');
title(['Error in numerical solution u:  index of u = ',...
       num2str(index_u)])
disp('Press any key to continue.'); pause

% Produce maximum errors over x as t varies.
maxerr_in_x = max(abs(error'));
figure; plot(t,maxerr_in_x); text(1.02*t_max,0,'t')
title('Maximum error for x in [0,1], as a function of t')

function true_u = true_soln(z,s)

switch index_u
    case 1
        true_u = s.^2 + z.^4;
```

```
        case 2
            true_u = exp(-0.1*s).*sin(pi*z);
end
end % true_u

function answer = G(z,s)

% This routine assumes s is a scalar, while z can be a vector.
switch index_u
        case 1
            answer = 2*s - 12*z.^2;
        case 2
            answer = (pi^2 - 0.1)*exp(-0.1*s).*sin(pi*z);
end
end % G

function answer = d0(s)
z = zeros(size(s));
answer = true_soln(z,s);
end % d0

function answer = d1(s)
z = ones(size(s));
answer = true_soln(z,s);
end % d1

function answer = f(z)
s = zeros(size(z));
answer = true_soln(z,s);
end % f


end % Test_MOL_Euler
```

Backward Euler method code:

```
function [x,t,u] = MOL_BEuler(d0,d1,f,G,T,h,m)
%
% function [x,t,u] = MOL_BEuler(d0,d1,f,G,T,h,m)
%
% Use the method of lines to solve
%     u_t = u_xx + G(x,t), 0 < x < 1, 0 < t < T
% with boundary conditions
%     u(0,t) = d0(t), u(1,t) = d1(t)
% and initial condition
%     u(x,0) = f(x).
% Use the backward Euler's method to solve the system of
```

```
% ODEs.  For the discretization, use a spatial stepsize of
% delta=1/m and a timestep of h.

x = linspace(0,1,m+1)'; delta = 1/m; delta_sqr = delta^2;
t = (0:h:T)'; N = length(t);

% Initialize u.
u = zeros(m+1,N);
u(:,1) = f(x);
u(1,:)  = d0(t); u(m+1,:)  = d1(t);

% Create tridiagonal coefficient matrix.
a = -(h/delta_sqr)*ones(m-1,1); c = a;
b = (1+2*h/delta_sqr)*ones(m-1,1);
a(1) = 0; c(m-1) = 0; option = 0;

% Solve for u using the backward Euler's method.
for n=2:N
      g = G(x(2:m),t(n));
      g(1) = g(1) + (1/delta_sqr)*u(1,n);
      g(m-1) = g(m-1) + (1/delta_sqr)*u(m+1,n);
      f = u(2:m,n-1) + h*g;
      switch option
      case 0     % first time:  factorize matrix
          [v,alpha,beta,message] = tridiag(a,b,c,f,m-1,option);
          option = 1;
      case 1     % other times:  use available factorization
          v = tridiag(alpha,beta,c,f,m-1,option);
      end
      u(2:m,n) = v;
end

u = u';
end % MOL_BEuler

function [x, alpha, beta, message] = tridiag(a,b,c,f,n,option)
%
% function [x, alpha, beta, message] = tridiag(a,b,c,f,n,option)
%
% Solve a tridiagonal linear system M*x=f
%
% INPUT:
% The order of the linear system is given as n.
% The subdiagonal, diagonal, and superdiagonal of M are given
% by the arrays a,b,c, respectively.  More precisely,
```

```
%      M(i,i-1) = a(i), i=2,...,n
%      M(i,i) = b(i), i=1,...,n
%      M(i,i+1) = c(i), i=1,...,n-1
% option=0 means that the original matrix M is given as
%      specified above.  We factorize M.
% option=1 means that the LU factorization of M is already
%      known and is stored in a,b,c.  This will have been
%      accomplished by a previous call to this routine.  In
%      that case, the vectors alpha and beta should have
%      been substituted for a and b in the calling sequence.
% All input values are unchanged on exit from the routine.
%
% OUTPUT:
% Upon exit, the LU factorization of M is already known and
% is stored in alpha,beta,c.  The solution x is given as well.
% message=0 means the program was completed satisfactorily.
% message=1 means that a zero pivot element was encountered
%      and the solution process was abandoned.  This case
%      happens only when option=0.

if option == 0
 alpha = a; beta = b;
 alpha(1) = 0;

 % Compute LU factorization of matrix M.
 for j=2:n
     if beta(j-1) == 0
         message = 1; return
     end
     alpha(j) = alpha(j)/beta(j-1);
     beta(j) = beta(j) - alpha(j)*c(j-1);
 end

 if beta(n) == 0
     message = 1; return
 end
end

% Compute solution x to M*x = f using LU factorization of M.
% Do forward substitution to solve lower triangular system.
if option == 1
    alpha = a; beta = b;
end
x = f; message = 0;
```

```
for j=2:n
    x(j) = x(j) - alpha(j)*x(j-1);
end
```

```
% Do backward substitution to solve upper triangular system.
x(n) = x(n)/beta(n);
for j=n-1:-1:1
    x(j) = (x(j) - c(j)*x(j+1))/beta(j);
end
```

```
end % tridiag
```

The test code for MOL_BEuler is essentially the same as that for MOL_Euler. In Test_MOL_Euler, simply replace the phrase MOL_Euler with MOL_BEuler throughout the code.

8.2 BACKWARD DIFFERENTIATION FORMULAS

The concept of a region of absolute stability is the initial tool used in studying the stability of a numerical method for solving stiff differential equations. We seek methods whose stability region each contains the entire negative real axis and as much of the left half of the complex plane as possible. There are a number of ways to develop such methods, but we discuss only one of them in this chapter — obtaining the *backward differentiation formulas* (BDFs).

Let $P_p(t)$ denote the polynomial of degree $\leq p$ that interpolates $Y(t)$ at the points $t_{n+1}, t_n, \ldots, t_{n-p+1}$ for some $p \geq 1$,

$$P_p(t) = \sum_{j=-1}^{p-1} Y(t_{n-j}) l_{j,n}(t), \tag{8.29}$$

where $\{l_{j,n}(t) : j = -1, \ldots, p-1\}$ are the Lagrange interpolation basis functions for the nodes $t_{n+1}, \ldots, t_{n-p+1}$ (see (B.4) in Appendix B). Use

$$P_p'(t_{n+1}) \approx Y'(t_{n+1}) = f(t_{n+1}, Y(t_{n+1})). \tag{8.30}$$

Combining (8.30) with (8.29) and solving for $Y(t_{n+1})$, we obtain

$$Y(t_{n+1}) \approx \sum_{j=0}^{p-1} \alpha_j Y(t_{n-j}) + h\beta f(t_{n+1}, Y(t_{n+1})). \tag{8.31}$$

The p-step BDF method is given by

$$y_{n+1} = \sum_{j=0}^{p-1} \alpha_j y_{n-j} + h\beta f(t_{n+1}, y_{n+1}). \tag{8.32}$$

Table 8.4 Coefficients of BDF method (8.32)

p	β	α_0	α_1	α_2	α_3	α_4	α_5
1	1	1					
2	$\frac{2}{3}$	$\frac{4}{3}$	$-\frac{1}{3}$				
3	$\frac{6}{11}$	$\frac{18}{11}$	$-\frac{9}{11}$	$\frac{2}{11}$			
4	$\frac{12}{25}$	$\frac{48}{25}$	$-\frac{36}{25}$	$\frac{16}{25}$	$-\frac{3}{25}$		
5	$\frac{60}{137}$	$\frac{300}{137}$	$-\frac{300}{137}$	$\frac{200}{137}$	$-\frac{75}{137}$	$\frac{12}{137}$	
6	$\frac{60}{147}$	$\frac{360}{147}$	$-\frac{450}{147}$	$\frac{400}{147}$	$-\frac{225}{147}$	$\frac{72}{147}$	$-\frac{10}{147}$

The coefficients for the cases of $p = 1, \ldots, 6$ are given in Table 8.4. The case $p = 1$ is simply the backward Euler method of (4.9) in Chapter 4. The truncation error for (8.32) can be obtained from the error formulas for numerical differentiation (e.g. see [11, (5.7.5)]),

$$T_n(Y) = -\frac{\beta}{p+1} h^{p+1} Y^{(p+1)}(\xi_n) \tag{8.33}$$

for some $t_{n-p+1} \leq \xi_n \leq t_{n+1}$.

The regions of absolute stability for the formulas of Table 8.4 are given in Figure 8.3. To create these regions, we must find all values $h\lambda$ for which

$$|r_j(h\lambda)| < 1, \qquad j = 0, 1, \ldots, p, \tag{8.34}$$

where the characteristic roots $r_j(h\lambda)$ are the solutions of

$$r^p = \sum_{j=0}^{p-1} \alpha_j r^{p-1-j} + h\lambda \beta r^p. \tag{8.35}$$

It can be shown that for $p = 1$ and $p = 2$, the BDF's are A-stable, and that for $3 \leq p \leq 6$, the region of absolute stability becomes smaller as p increases, although containing the entire negative real axis in each case. For $p \geq 7$, the regions of absolute stability are not acceptable for the solution of stiff problems. This is discussed in greater detail in the following section.

8.3 STABILITY REGIONS FOR MULTISTEP METHODS

Recalling (7.1), all general multistep methods, including AB, AM, and BDF (and other) methods, can be represented as follows:

$$y_{n+1} = \sum_{j=0}^{p} a_j y_{n-j} + h \sum_{j=-1}^{p} b_j f(t_{n-j}, y_{n-j}). \tag{8.36}$$

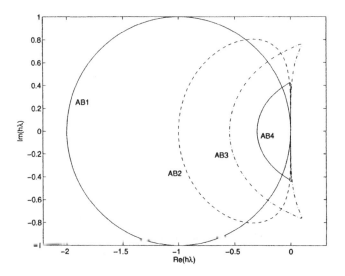

Figure 8.1   Stability regions for Adams–Bashforth methods. Note that AB1 is Euler's method

For the test equation $dY/dt = \lambda Y$, we have $f(t, Y) = \lambda Y$; and recalling (7.42), the characteristic polynomial for (8.36) is

$$0 = (1 - h\lambda b_{-1}) r^{p+1} - \sum_{j=0}^{p} (a_j + h\lambda b_j) r^{p-j}. \tag{8.37}$$

The *boundary* of the stability region is where all roots of this characteristic equation have magnitude 1 or less, and at least one root with magnitude 1. We can find all the values of $h\lambda$ where one of the roots has magnitude 1. All roots with magnitude 1 can be represented as $r = e^{i\theta}$ with $i = \sqrt{-1}$. So we can find all $h\lambda$ where (8.37) holds with $r = e^{i\theta}$. Separating out $h\lambda$ gives

$$r^{p+1} - \sum_{j=0}^{p} a_j r^{p-j} = h\lambda \sum_{j=-1}^{p} b_j r^{p-j},$$

$$h\lambda = \left(r^{p+1} - \sum_{j=0}^{p} a_j r^{p-j} \right) \div \left(\sum_{j=-1}^{p} b_j r^{p-j} \right),$$

where $r = e^{i\theta}$ for $0 \le \theta \le 2\pi$ gives a set that includes the boundary of the stability region. With a little more care, we can identify which of the regions separated by this curve form the true stability region.

Remark. From Section 7.3 of Chapter 7, the root condition (7.32)–(7.33) is necessary for convergence and stability of a multistep method. This form of stability

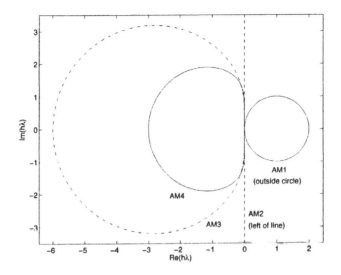

Figure 8.2   Stability regions for Adams–Moulton methods. Note that AM1 is the implicit Euler method, and AM2 is the trapezoidal method. Note the different scale on the axes as compared to Figure 8.1

is sometimes also called *weak stability*, as we ordinarily require additional stability conditions for a practical numerical method. Without the root condition, the method cannot be expected to produce numerical solutions that approach the true solution as $h \to 0$, regardless of the value of λ. The root condition sometimes fails for certain consistent multistep methods, but almost no one discusses those methods because they are useless except to explain the importance of stability! As a simple example of such a method, recall Example 7.34 from Section 7.3.

8.4 ADDITIONAL SOURCES OF DIFFICULTY

8.4.1 A-stability and L-stability

There are still problems with the BDF methods and with other methods that are chosen solely on the basis of their region of absolute stability. First, with the model equation $Y' = \lambda Y$, if $\mathrm{Real}(\lambda)$ is of large magnitude and negative, then the solution $Y(t)$ goes to the zero very rapidly, and as $\mathrm{Real}(\lambda) \to -\infty$, the convergence to zero of $Y(t)$ becomes more rapid. We would like the same behavior to hold for the numerical solution of the model equation $\{y_n\}$. To illustrate this idea, we show that the A-stable trapezoidal rule does not maintain this behavior.

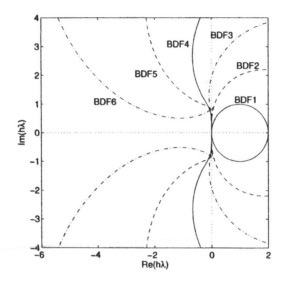

Figure 8.3   Stability regions for backward difference formula methods. Note that BDF1 is again the implicit Euler method. The labels are inside the stability region for the labeled method.

Apply the trapezoidal method (4.22) to the model equation (8.1). Doing so leads to the numerical approximation

$$y_n = \left[\frac{1 + \frac{1}{2}h\lambda}{1 - \frac{1}{2}h\lambda} \right]^n, \qquad n \geq 0. \tag{8.38}$$

If $|\text{Real}(\lambda)|$ is large, then the fraction inside the brackets is less than 1 in magnitude, but is nearly equal to -1; and thus y_n decreases to 0 quite slowly. This suggests that the trapezoidal method may not be a completely satisfactory choice for stiff problems.

In comparison, the A-stable backward Euler method has the desired behavior. From (4.10) in Chapter 4, the solution of the model problem is

$$y_n = \left[\frac{1}{1 - h\lambda} \right]^n, \qquad n \geq 0.$$

As $|\lambda|$ increases, the sequence $\{y_n\}$ goes to zero more rapidly. Thus the backward Euler solution better reflects the behavior of the true solution of the model equation. An A-stable numerical method is called *L-stable* if at each fixed $t = t_n$, the numerical solution y_n at t_n satisfies $y_n \to 0$ as $\text{Real}(\lambda) \to -\infty$. The trapezoidal rule is not L-stable, whereas the backward Euler method is L-stable. This material was explored earlier in Problems 14 and 15 of Chapter 4.

8.4.2 Time-varying problems and stability

A second problem with the use of stability regions to determine methods for stiff problems is that it is based on using constant λ and linear problems. The linearization (8.10) is often valid, but not always. For example, consider the second-order linear problem

$$y'' + ay' + (1 + b \cdot \cos(2\pi y))y = g(t), \qquad t \geq 0, \tag{8.39}$$

in which one coefficient is not constant. Convert it to the equivalent system

$$\begin{aligned} y_1' &= y_2, \\ y_2' &= -(1 + b \cdot \cos(2\pi t))y_1 - ay_2 + g(t). \end{aligned} \tag{8.40}$$

We assume $a > 0$, $|b| < 1$. The eigenvalues of the Jacobian matrix for this system are

$$\lambda = \frac{-a \pm \sqrt{a^2 - 4[1 + b \cdot \cos(2\pi t)]}}{2}. \tag{8.41}$$

These are either negative real numbers or complex numbers with negative real parts. On the basis of the stability theory for the constant coefficient (or constant Λ) case, we might be led to assume that the effect of all perturbations in the initial data for (8.40) would die away as $t \to \infty$. But in fact, the homogeneous part of (8.39) will have unbounded solutions. Thus there will be perturbations of the initial values that will lead to unbounded perturbed solutions in (8.39). This calls into question the validity of the use of the model equation $y' = \lambda y + g(t)$. Using the model equation (8.1) suggests methods that we may want to study further; but by itself, this approach is not sufficient to encompass the vast variety of linear and nonlinear problems. The example (8.39) is taken from Aiken [1, p. 269].

8.5 SOLVING THE FINITE-DIFFERENCE METHOD

We illustrate the difficulty in solving the finite difference equations by considering the backward Euler method,

$$y_{n+1} = y_n + hf(t_{n+1}, y_{n+1}), \qquad n \geq 0 \tag{8.42}$$

first for a single equation and then for a system of equations. For a single equation, we summarize the discussion involving (4.12)–(4.16) of Chapter 4. If the ordinary iteration formula

$$y_{n+1}^{(j+1)} = y_n + hf(t_{n+1}, y_{n+1}^{(j)}), \qquad j \geq 0 \tag{8.43}$$

is used, then

$$y_{n+1} - y_{n+1}^{(j+1)} \approx h \frac{\partial f(t_{n+1}, y_{n+1})}{\partial y} \left[y_{n+1} - y_{n+1}^{(j)} \right].$$

For convergence, we would need to have

$$\left| h \frac{\partial f(t_{n+1}, y_{n+1})}{\partial y} \right| < 1. \tag{8.44}$$

But with stiff equations, this would again force h to be very small, which we are trying to avoid. Thus another rootfinding method must be used to solve for y_{n+1} in (8.42).

The most popular methods for solving (8.42) are based on Newton's method and variants of it. For a single differential equation, Newton's method for finding y_{n+1} is

$$
\begin{aligned}
y_{n+1}^{(j+1)} = y_{n+1}^{(j)} &- \left[1 - h f_y(t_{n+1}, y_{n+1}^{(j)}) \right]^{-1} \\
&\times \left[y_{n+1}^{(j)} - y_n - h f(t_{n+1}, y_{n+1}^{(j)}) \right]
\end{aligned}
\tag{8.45}
$$

for $j \geq 0$. A crude initial guess is $y_{n+1}^{(0)} = y_n$, although generally this can be improved on.

With a system of m differential equations, as in (8.9), Newton's method is

$$
\begin{aligned}
\left[I - h\mathbf{f}_y(t_{n+1}, \mathbf{y}_{n+1}^{(j)}) \right] \delta_n^{(j)} &= \mathbf{y}_{n+1}^{(j)} - \mathbf{y}_n - h\mathbf{f}(t_{n+1}, \mathbf{y}_{n+1}^{(j)}), \\
\mathbf{y}_{n+1}^{(j+1)} &= \mathbf{y}_{n+1}^{(j)} - \delta_n^{(j)}, \qquad j \geq 0.
\end{aligned}
\tag{8.46}
$$

This is a system of m linear simultaneous equations for the vector $\delta_n^{(j)} \in \mathbb{R}^m$, and such a linear system must be solved repeatedly at each step t_n. The matrix of coefficients changes with each iterate $\mathbf{y}_{n+1}^{(j)}$ and with each step t_n. This rootfinding procedure is usually costly to implement; consequently, we seek variants of Newton's method that require less computation time.

As one approach to decreasing the cost of (8.46), the matrix approximation

$$
I - h\mathbf{f}_y(t_{n+1}, \mathbf{z}) \approx I - h\mathbf{f}_y(t_{n+1}, \mathbf{y}_{n+1}^{(j)}), \qquad \text{some } \mathbf{z} \approx \mathbf{y}_n
\tag{8.47}
$$

is used for all j and for a number of successive steps t_n. Thus Newton's method (8.46) is approximated by

$$
\begin{aligned}
\left[I - h\mathbf{f}_y(t_{n+1}, \mathbf{z}) \right] \delta_n^{(j)} &= \mathbf{y}_{n+1}^{(j)} - \mathbf{y}_n - h\mathbf{f}\left(t_{n+1}, \mathbf{y}_{n+1}^{(j)} \right), \\
\mathbf{y}_{n+1}^{(j+1)} &= \mathbf{y}_{n+1}^{(j)} - \delta_n^{(j)}, \qquad j \geq 0.
\end{aligned}
\tag{8.48}
$$

This amounts to solving a number of linear systems with the same coefficient matrix. This can be done much more cheaply than when the matrix is being modified with each iteration and each new step t_n. The matrix in (8.47) will have to be updated periodically, but the savings will still be very significant when compared to an exact Newton method. For a further discussion of this topic, see Aiken [1, p. 7].

8.6 COMPUTER CODES

The MATLAB program ode15s is used to solve stiff ordinary differential equations. It is based on a modification of the variable order family of BDF methods discussed earlier in the chapter. Details of the actual methods and their implementation can be found in Shampine and Reichelt [73, Section 2]. The nonlinear finite difference system (see (8.42) for the backward Euler method) at each timestep t_n is solved by a variant of the modified Newton method of (8.47)–(8.48). The program ode15s is used in precisely the same manner as the program ode45 discussed in Chapter 5 and the program ode113 of Chapter 6; and the entire suite of MATLAB ODE programs is discussed at length in [73], [74].

A package of programs called *Sundials* [46] includes state-of-the-art programs for solving initial value problems for ordinary differential equations, including stiff equations, and differential algebraic equations. Included is an interface for use with MATLAB. The *Sundials* package is the latest in a sequence of excellent programs from the national energy laboratories (especially Lawrence-Livermore Laboratory and Sandia Laboratory) in the USA, for use in solving ordinary differential equations and developed over more than 30 years.

A general presentation of the method of lines is given in Schiesser [71]. For some older "method of lines codes" to solve systems of nonlinear parabolic partial differential equations in one and two space variables, see Sincovec and Madsen [75] and Melgaard and Sincovec [63]. For use with MATLAB, the Partial Differential Equations Toolbox solves partial differential equations, and it contains a \method of lines codes" code to solve parabolic equations. It also makes use of the MATLAB suite of programs for solving ordinary differential equations.

Example 8.4 We modify the program test_ode45 of Section 5.5 by replacing ode45 with ode15s throughout the code. We illustrate the use of ode15s with the earlier example (8.8), solving it on $[0, 20]$ and using $AbsTol = 10^{-6}$, $RelTol = 10^{-4}$. We choose λ to be negative, but allow it to have a large magnitude, as in Example 8.2 for the Adams–Bashforth method of order 2 (see Table 8.1). As a comparison to ode15s, we also give the results obtained using ode45 and ode113. We give the number of needed derivative evaluations with the three programs, and we also give the maximum error in the computed solution over $[0, 20]$. This maximum error is for the interpolated solution at the points defined in the test program test_ode15s. The results, shown in Table 8.5, indicate clearly that as the stiffness increases (or as $|\lambda|$ increases), the efficiencies of ode45 and ode113 decreases. In comparison, the code ode15s is relatively unaffected by the increasing magnitude of $|\lambda|$. ■

PROBLEMS

1. Derive the BDF method of order 2.

2. Consider the BDF method of order 2. Show that its region of absolute stability contains the negative real axis, $-\infty < h\lambda < 0$.

3. Using the BDF method of order 2, repeat the calculations in Example 8.2. Comment on your results.
   Hint: Note that the linearity of the test equation (8.8) allows the implicit BDF equation for y_{n+1} to be solved explicitly; iteration is unnecessary.

Table 8.5 Comparison of ode15s, ode45, and ode113 for the stiff equation (8.8)

	ode15s	ode45	ode113
$\lambda = -1$			
Maximum error	$5.44e - 4$	$1.43e - 4$	$3.40e - 4$
Function evaluations	235	229	132
$\lambda = -10$			
Maximum error	$1.54e - 4$	$4.51e - 5$	$9.05e - 4$
Function evaluations	273	979	337
$\lambda = -50$			
Maximum error	$8.43e - 5$	$4.24e - 5$	$1.41e - 3$
Function evaluations	301	2797	1399
$\lambda = -500$			
Maximum error	$4.67e - 5$	$1.23e - 4$	$3.44e - 3$
Function evaluations	309	19663	13297

4. Implement MOL_Euler. Use it to experiment with various choices of δ and h with the true solution $U = e^{-0.1t}\sin(\pi x)$. Use some values of δ and h that satisfy (8.24) and others not satisfying it. Comment on your results.

5. Implement MOL_Euler and MOL_BEuler. Experiment as in Example 8.3. Use various values of h and δ. Do so for the following true solutions U (note that the functions d_0, d_1, f, and G are determined from the known test case U):

   (a) $U = x^4 + t^2$.
   (b) $U = (1 - e^{-t})\cos(\pi x)$.
   (c) $U = \exp(1/(t+1))\cos(\pi x)$.

CHAPTER 9

IMPLICIT RK METHODS FOR STIFF DIFFERENTIAL EQUATIONS

Runge–Kutta methods were introduced in Chapter 5, and we now want to consider them as a means of solving stiff differential equations. When working with multistep methods in Chapter 8, we needed to use implicit methods in order to solve stiff equations; the same is true with Runge–Kutta methods. Also, as with multistep methods, we need to develop the appropriate stability theory and carefully analyze what happens when we apply these methods to stiff equations.

9.1 FAMILIES OF IMPLICIT RUNGE–KUTTA METHODS

Runge–Kutta methods can be used for stiff differential equations. However, we need *implicit* Runge–Kutta methods, which were introduced in Section 5.6 of Chapter 5.

149

The general forms of these equations, for a method with s stages, are as follows:

$$z_{n,i} = y_n + h \sum_{j=1}^{s} a_{i,j} f(t_n + c_j h,\, z_{n,j}), \qquad i = 1, \ldots, s, \qquad (9.1)$$

$$y_{n+1} = y_n + h \sum_{j=1}^{s} b_j f(t_n + c_j h,\, z_{n,j}). \qquad (9.2)$$

Note that the equation for $z_{n,i}$ involves *all* the $z_{n,j}$ values. So for implicit Runge–Kutta methods we need to solve an extended system of equations. If each y_n is a real number, then we have a system of s equations in s unknowns for each timestep. If each y_n is a vector of dimension N then we have a system of Ns equations in Ns unknowns. As in Chapter 5, we can represent implicit Runge–Kutta methods in terms of Butcher tableaus (see (5.26)),

$$
\begin{array}{c|ccccc}
c_1 & a_{1,1} & a_{1,2} & \cdots & a_{1,s-1} & a_{1,s} \\
c_2 & a_{2,1} & a_{2,1} & \cdots & a_{2,s-1} & a_{2,s} \\
c_3 & a_{3,1} & a_{3,2} & \cdots & a_{3,s-1} & a_{3,s} \\
\vdots & \vdots & \vdots & \ddots & \vdots & \vdots \\
c_s & a_{s,1} & a_{s,2} & \cdots & a_{s,s-1} & a_{s,s} \\
\hline
& b_1 & b_2 & \cdots & b_{s-1} & b_s
\end{array}
\qquad \text{or} \qquad
\begin{array}{c|c}
\mathbf{c} & A \\
\hline
& \mathbf{b}^T
\end{array}
\qquad (9.3)
$$

Some implicit methods we have already seen are actually implicit Runge–Kutta methods, namely, the backward Euler method and the trapezoidal rule. Their Butcher tableaus are shown in Tables 9.1 and 9.2.

Table 9.1   Butcher tableau - backward Euler method

$$
\begin{array}{c|c}
1 & 1 \\
\hline
& 1
\end{array}
$$

Table 9.2   Butcher tableau - trapezoidal method

$$
\begin{array}{c|cc}
0 & 0 & 0 \\
1 & 1/2 & 1/2 \\
\hline
& 1/2 & 1/2
\end{array}
$$

These methods are also BDF methods. However, higher-order BDF methods require y_{n-1} to compute y_{n+1} and so they are not Runge–Kutta methods.

Higher-order Runge–Kutta methods have been developed, although the conditions that need to be satisfied for such Runge–Kutta methods to have order p become very complex for large p. Nevertheless, a few families of Runge–Kutta methods with arbitrarily high-order accuracy have been created. One such family is the set of

Gauss methods given in (5.63)–(5.64) of Chapter 5; they are closely related to Gauss–Legendre quadrature for approximating integrals. These have the property that the c_i values are the roots of the Legendre polynomial

$$\frac{d^s}{dx^s} \left[x^s \left(1 - x \right)^s \right].$$

The other coefficients of these methods can be determined from the c_i values by means of the so-called *simplifying assumptions* of Butcher [23]:

$$B(p): \quad \sum_{i=1}^{s} b_i c_i^{k-1} = \frac{1}{k}, \quad k = 1, 2, \ldots, p, \tag{9.4}$$

$$C(q): \quad \sum_{j=1}^{s} a_{ij} c_j^{k-1} = \frac{c_i^k}{k}, \quad k = 1, 2, \ldots, q, \quad i = 1, 2, \ldots, s, \tag{9.5}$$

$$D(r): \quad \sum_{i=1}^{s} b_i c_i^{k-1} a_{ij} = \frac{b_j}{k} \left(1 - c_k^k \right),$$

$$k = 1, 2, \ldots, r, \quad j = 1, 2, \ldots, s. \tag{9.6}$$

Condition $B(p)$ says that the quadrature formula

$$\int_t^{t+h} f(s)\, ds \approx h \sum_{i=1}^{s} b_i\, f(t + c_j h)$$

is exact for all polynomials of degree $< p$. If this condition is satisfied, we say that the Runge–Kutta method has *quadrature order* p. Condition $C(q)$ says that the corresponding quadrature formulas on $[t, t + c_i h]$, namely

$$\int_t^{t+c_i h} f(s)\, ds \approx h \sum_{j=1}^{s} a_{ij}\, f(t + c_j h)$$

are exact for all polynomials of degree $< q$. If this condition is satisfied, we say that the Runge–Kutta method has *stage order* q. The importance of these assumptions is demonstrated in the following theorem of Butcher [23, Thm. 7].

Theorem 9.1 *If a Runge–Kutta method satisfies conditions $B(p)$, $C(q)$, and $D(r)$ with $p \leq q + r + 1$ and $p \leq 2q + 2$, its order of accuracy is p.*

We can use this theorem to construct the Gauss methods. First we choose $\{c_i\}$, the quadrature points of the Gaussian quadrature. This can be done by looking up tables of these numbers, and then scaling and shifting them from the interval $[-1, +1]$ to $[0, 1]$. Alternatively, they can be computed as zeros of appropriate Legendre polynomials [11, Section 5.3]. We then choose the quadrature weights b_i to make $B(p)$ true for as large a value of p as possible. For the Gaussian quadrature points, this is $p = 2s$. Note that if condition $B(p)$ fails, then the method *cannot* have order p.

This leaves us with the s^2 coefficients a_{ij} to find. These can be determined by applying conditions $C(q)$ and $D(r)$ with sufficiently large q and r. Fortunately, there are some additional relationships between these conditions. It turns out that if $B(q+r)$ and $C(q)$ hold, then $D(r)$ holds as well. Also if $B(q+r)$ and $D(r)$ hold, then so does $C(q)$ [23, Thms. 3, 4, 5 & 6].

So we just need to satisfy $C(s)$ in addition. Then $B(2s)$ and $C(s)$ together imply $D(s)$; setting $q = r = s$ and $p = 2s$ in Theorem 9.1 gives us a method of order $2s$. Imposing condition $C(s)$ gives us exactly s^2 linear equations for the a_{ij} values, which can be easily solved. Thus the order of the s-stage Gauss method is $2s$.

Some Gauss methods are shown in Tables 9.3–9.5. For the derivation of these formulas, refer back to Section 5.6 in Chapter 5.2. The two-point Gauss method was given in (5.70)-(5.71) of Section 5.6.1.

Table 9.3   Butcher tableau for Gauss method of order 2

$$
\begin{array}{c|c}
1/2 & 1/2 \\
\hline
 & 1
\end{array}
$$

Table 9.4   Butcher tableau for Gauss method of order 4

$$
\begin{array}{c|cc}
\left(3-\sqrt{3}\right)/6 & 1/4 & \left(3-2\sqrt{3}\right)/12 \\
\left(3+\sqrt{3}\right)/6 & \left(3+2\sqrt{3}\right)/12 & 1/4 \\
\hline
 & 1/2 & 1/2
\end{array}
$$

Table 9.5   Butcher tableau for Gauss method of order 6

$$
\begin{array}{c|ccc}
\left(5-\sqrt{15}\right)/10 & 5/36 & 2/9-\sqrt{15}/5 & 5/36-\sqrt{15}/30 \\
1/2 & 5/36+\sqrt{15}/24 & 2/9 & 5/36-\sqrt{15}/24 \\
\left(5+\sqrt{15}\right)/10 & 5/36+\sqrt{15}/30 & 2/9+\sqrt{15}/5 & 5/36 \\
\hline
 & 5/18 & 4/9 & 5/18
\end{array}
$$

There are some issues that Gauss methods do not address, and so a number of closely related methods have been developed. The most important of these are the Radau methods, particularly the Radau IIA methods. For the Radau IIA methods the c_i terms are roots of the polynomial

$$
\frac{d^{s-1}}{dx^{s-1}}\left[x^{s-1}\left(1-x\right)^s\right].
$$

In particular, we have $c_s = 1$, as we can see in Tables 9.6 and 9.7, which show the lower-order Radau IIA methods. The simplifying assumptions satisfied by the Radau IIA methods are $B(2s-1)$, $C(s)$, and $D(s-1)$, so that the order of a Radau IIA method is $2s-1$. The order 1 Radau IIA method is just the implicit Euler method, given in Table 9.1. The derivation of these formulas is similar to that for the Gauss formulas, only now we are using Radau quadrature rules rather than Gauss–Legendre quadrature rules; see Section 5.6.

Table 9.6 Butcher tableau for Radau method of order 3

$$
\begin{array}{c|cc}
1/3 & 5/12 & -1/12 \\
1 & 3/4 & 1/4 \\
\hline
 & 3/4 & 1/4
\end{array}
$$

Table 9.7 Butcher tableau for Radau method of order 5

$$
\begin{array}{c|ccc}
\left(4 - \sqrt{6}\right)/10 & \left(88 - 7\sqrt{6}\right)/360 & \left(296 - 169\sqrt{6}\right)/1800 & \left(-2 + 3\sqrt{6}\right)/225 \\
\left(4 + \sqrt{6}\right)/10 & \left(296 + 169\sqrt{6}\right)/1800 & \left(88 + 7\sqrt{6}\right)/360 & \left(-2 - 3\sqrt{6}\right)/225 \\
1 & \left(16 - \sqrt{6}\right)/36 & \left(16 + \sqrt{6}\right)/36 & 1/9 \\
\hline
 & \left(16 - \sqrt{6}\right)/36 & \left(16 + \sqrt{6}\right)/36 & 1/9
\end{array}
$$

A third family of Runge–Kutta methods worth considering are the Lobatto IIIC methods; the c_j values are the roots of the polynomial

$$
\frac{d^{s-2}}{dx^{s-2}} \left[x^{s-1}(1 - x)^{s-1} \right],
$$

and we use the simplifying conditions $B(2s - 2)$, $C(s - 1)$, and $D(s - 1)$. The Lobatto IIIC methods have $c_1 = 0$ and $c_s = 1$. The order of the s-stage Lobatto IIIC method is $2s - 2$.

Other Runge–Kutta methods have been developed to handle various other issues. For example, while general implicit Runge–Kutta methods with s stages require the solution of a system of Ns equations in Ns unknowns, some implicit Runge–Kutta methods require the solution of a sequence of s systems of N equations in N unknowns. This is often simpler than solving Ns equations in Ns unknowns. These methods are known as *diagonally implicit Runge–Kutta methods* (DIRK methods). For these methods we take $a_{i,j} = 0$ whenever $i < j$. Two examples of DIRKs are given in Table 9.8. The method of Alexander [2] is an order 3 method with three stages. The method of Crouzeix and Raviart [31] is an order 4 method with three stages. The constants in Alexander's method are

$$
\alpha = \text{the root of } \quad x^3 - 3x^2 + \tfrac{3}{2}x - \tfrac{1}{6} \ \text{ in } \ (\tfrac{1}{6}, \tfrac{1}{2}),
$$
$$
\tau_2 = \tfrac{1}{2}(1 + \alpha),
$$
$$
b_1 = -\tfrac{1}{4}(6\alpha^2 - 16\alpha + 1),
$$
$$
b_2 = \tfrac{1}{4}(6\alpha^2 - 20\alpha + 5).
$$

The constants in Crouzeix and Raviart's method are given by

$$\gamma = \frac{1}{\sqrt{3}} \cos\left(\frac{\pi}{18}\right) + \frac{1}{2},$$

$$\delta = \frac{1}{6 \left(2\gamma - 1\right)^2}.$$

There are a large number of DIRK methods, and some of them can be found, for example, in Hairer and Wanner's text [44].

Table 9.8   Butcher tableau for DIRK methods

α	α		
τ_2	$\tau_2 - \alpha$	α	
1	b_1	b_2	α
	b_1	b_2	α

(a) Method of Alexander

γ	γ		
$1/2$	$1/2 - \gamma$	γ	
$1 - \gamma$	2γ	$1 - 4\gamma$	γ
	δ	$1 - 2\delta$	δ

(b) Method of Crouzeix & Raviart

9.2 STABILITY OF RUNGE–KUTTA METHODS

Implicit Runge–Kutta methods need the same kind of stability properties as found in multistep methods if they are to be useful in solving stiff differential equations. Fortunately, most of the stability aspects can be derived using some straightforward linear algebra.

Consider the model differential equation

$$Y' = \lambda Y.$$

Following (9.1)–(9.2), denote $\mathbf{z}_n^T = [z_{n,1}, z_{n,2}, \ldots, z_{n,s}]$. Apply (9.1)–(9.2) to this differential equation:

$$\mathbf{z}_n = y_n \, \mathbf{e} + h\lambda A \, \mathbf{z}_n,$$

$$y_{n+1} = y_n + h\lambda \mathbf{b}^T \mathbf{z}_n.$$

Here $\mathbf{e}^T = [1, 1, \ldots, 1]$ is the s-dimensional vector of all ones. Some easy algebra gives

$$y_{n+1} = \left[1 + h\lambda \mathbf{b}^T \left(I - h\lambda A\right)^{-1} \mathbf{e}\right] y_n = R(h\lambda) \, y_n.$$

The stability function is

$$R(\eta) = 1 + \eta \, \mathbf{b}^T \left(I - \eta A\right)^{-1} \mathbf{e}. \tag{9.7}$$

As before, the Runge–Kutta method is A-stable if $|R(\eta)| < 1$ for all complex η with Real $\eta < 0$.

All Gauss (Tables 9.3-9.5), Radau IIA (Tables 9.1, 9.6, 9.7), and some DIRK methods (Table 9.8) are A-stable, which makes them stable for *any* λ with Real $\lambda < 0$. However, this does not necessarily make them *accurate*. For more on this topic, see the following section on order reduction.

For nonlinear problems, there is another form of stability that is very useful, called *B-stability*. This is based on differential equations

$$Y' = f(t, Y), \qquad Y(t_0) = Y_0,$$

where $f(t, y)$ satisfies only a *one-sided Lipschitz condition*:

$$(y - z)^T (f(t, y) - f(t, z)) \le \mu \|y - z\|^2.$$

If $f(t, y)$ is Lipschitz in y with Lipschitz constant L (see (1.10) in Chapter 1), then it automatically satisfies the one-sided Lipschitz condition with $\mu = L$. However, the reverse need not hold. For example, the system of differential equations (8.16) obtained for the heat equation in Section 8.1 satisfies the one-sided Lipschitz condition with $\mu = 0$, no matter how fine the discretization. The ordinary Lipschitz constant, however, is roughly proportional to m^2, where m is the number of grid points chosen for the space discretization.

The importance of one-sided Lipschitz conditions is that they are closely related to stability of the differential equation. In particular, if

$$Y' = f(t, Y), \qquad Y(t_0) = Y_0,$$
$$Z' = f(t, Z), \qquad Z(t_0) = Z_0,$$

and $f(t, y)$ satisfies the one-sided Lipschitz condition with constant μ, then

$$\|Y(t) - Z(t)\| \le e^{\mu(t - t_0)} \|Y_0 - Z_0\|.$$

This can be seen by differentiating

$$m(t) = \|Y(t) - Z(t)\|^2 = (Y(t) - Z(t))^T (Y(t) - Z(t))$$

as follows:

$$m'(t) = 2 (Y(t) - Z(t))^T (Y'(t) - Z'(t))$$
$$= 2 (Y(t) - Z(t))^T [f(t, Y(t)) - f(t, Z(t))]$$
$$\le 2\mu \|Y(t) - Z(t)\|^2 = 2\mu m(t).$$

Hence

$$m(t) \le e^{2\mu(t - t_0)} m(t_0),$$

and taking square roots gives

$$\|Y(t) - Z(t)\| \le e^{\mu(t - t_0)} \|Y_0 - Z_0\|.$$

The case where the one-sided Lipschitz constant μ is zero means that the differential equation is *contractive*; that is, different solutions cannot become further apart or separated. If we require that the numerical solution be also contractive $(\|y_{n+1} - z_{n+1}\| \leq \|y_n - z_n\|$ for any two numerical solutions y_k and $z_k)$ whenever $\mu = 0$, then the method is called *B-stable* [24]. This condition seems very useful, but rather difficult to check. Fortunately, a simple and easy condition to test was found independently in [22] and [30]: namely, if

$$b_i \geq 0 \quad \text{for all } i \tag{9.8}$$

and

$$M = [b_i a_{ij} + b_j a_{ji} - b_i b_j]_{i,j=1}^s \text{ is positive semidefinite} \tag{9.9}$$

(i.e., $w^T M w \geq 0$ for all vectors w), then the Runge–Kutta method is B-stable. Testing a matrix M for being positive semidefinite is actually quite easy. One test is to compute the eigenvalues of M if M is symmetric. If all eigenvalues are ≥ 0, then M is positive semidefinite. For a nonsymmetric matrix M, it is positive semidefinite if all the eigenvalues of the matrix $(M + M^T)/2$ are nonnegative.

If a method is B-stable, then it is A-stable. To see this, for a B-stable method we can look at the differential equation

$$Y' = \begin{bmatrix} \alpha & +\beta \\ -\beta & \alpha \end{bmatrix} Y,$$

which has the one-sided Lipschitz constant $\mu = 0$ if $\alpha \leq 0$. The eigenvalues of this 2×2 matrix are $\alpha \pm i\beta$, which are in the left half of the complex plane if $\alpha < 0$. So if a method is B-stable, then $\alpha \leq 0$ implies that the numerical solution is contractive, and thus the stability region includes the left half-plane; that is, the method is A-stable.

This test for B-stability quickly leads to the realization that a number of important families of implicit Runge–Kutta methods are B-stable, such as the Gauss methods, the Radau IA, and the Radau IIA methods. The DIRK method in Table 9.8 (part b) is, however, A-stable but not B-stable. What does this mean in practice? For strongly nonlinear problems, A-stability may not suffice to ensure good behavior of the numerical method, especially if we consider integration for long time periods. It also means that Gauss or Radau IIA methods are probably better than DIRK methods despite the extra computational cost of the Gauss and Radau methods.

9.3 ORDER REDUCTION

Stability is clearly necessary, but it is not sufficient to obtain accurate solutions to stiff systems of ordinary differential equations. A phenomenon that is commonly observed is that when applied to stiff problems, many implicit methods do not seem to achieve the order of accuracy that is expected for the method. This phenomenon is called *order reduction* [44, pp. 225–228].

Order reduction occurs for certain Runge–Kutta methods, but not for BDF methods.

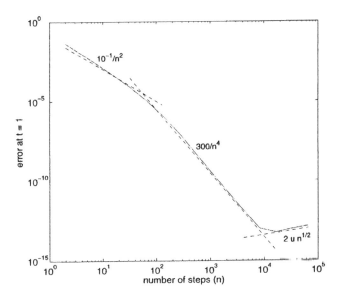

Figure 9.1   Error norms for the test equation (9.10)

Example 9.2   Consider, for example, the fourth-order Gauss method with $s = 2$ (see Tables 9.3–9.5). Figure 9.1 shows how the error behaves for a test equation

$$Y' = D\,(Y - g(t)) + g'(t), \qquad Y(0) = g(0). \tag{9.10}$$

For this particular example, D is a 100×100 diagonal matrix with negative diagonals randomly generated in the range from -2^{-20} to $-2^{+20} \approx -10^6$. The diagonal entries are exponentials of uniformly distributed pseudo-random values. The function $g(t)$ likewise involves pseudo-random numbers, but is a smooth function of t. The exact solution is $Y(t) = g(t)$, so we can easily compute errors in the numerical solution. For the function $g(t)$ we used $g(t) = \cos(t)\,z_1 - \exp(-t)\,z_2$ with z_1, z_2 randomly generated vectors using a normal distribution.

Note that the Gauss method with $s = 2$ is a fourth-order method, so that we expect the errors to be $\mathcal{O}(h^4)$ as the stepsize becomes small. But this ignores two factors: (1) the hidden constant in the $\mathcal{O}$ expression may be quite large because of the stiffness of the differential equation, and (2) asymptotic results like this are true provided h is "small enough". How small is "small enough" depends on the problem, and for stiff differential equations, this can depend on how stiff the equation is. Make the stiffness go to infinity, and the limit for "small enough" may go to zero. If that happens, then the standard convergence theory may be meaningless for practical stiff problems.

As can be seen from Figure 9.1, the error for larger values of h seems to behave more like $\mathcal{O}(h^2)$ than $\mathcal{O}(h^4)$. Also, for smaller values of h we see $\mathcal{O}(h^4)$ error behavior (as we might expect), but with a large value for the hidden constant inside the $\mathcal{O}$. For very small h and many steps, we see that roundoff error from floating-point arithmetic limits the accuracy possible with this method. (The quantity u in the graph denotes the unit round of the floating-point arithmetic.) If we increase the stiffness of the problem as we reduce the size of h, we might only see the $\mathcal{O}(h^2)$ behavior of the error. This is the effect of *order reduction*. ∎

Order reduction can be explained in terms of the following simple version of the test differential equation (9.10),

$$Y'(t) = \lambda(Y - g(t)) + g'(t), \qquad Y(t_0) = g(t_0). \qquad (9.11)$$

The exact solution is $Y(t) = g(t)$ for all t. However, the numerical solution of this is not exact, particularly if $h\lambda$ is large. What we want to find out is the magnitude of the error in terms of h *independently of* λh. This can be different from the order of the error for fixed λ as $h \to 0$. The Runge–Kutta equations are

$$z_{n,i} = y_n + h\sum_{j=1}^{s} a_{ij}\, f(t_n + c_j h, z_{n,j}), \qquad i = 1, 2, \ldots, s.$$

From this formula, it seems that the intention is for $z_{n,i} \approx Y(t_n + c_i h)$. Consider for a moment the even simpler test problem

$$\frac{dY}{dt} = g'(t), \qquad Y(t_0) = g(t_0).$$

The *stage order* of a Runge–Kutta method comes from the order of the error in the approximation $z_{n,i} \approx g(t_n + c_i h)$,

$$g(t_n + c_i h) = g(t_n) + h\sum_{j=1}^{s} a_{ij}\, g'(t_n + c_j h) + \mathcal{O}(h^{q+1})$$

for all i, indicating a stage order of q. The *quadrature order* is the order of the final formula for this very simple test equation; the result

$$g(t_n + h) = g(t_n) + h\sum_{j=1}^{s} b_j\, g'(t_n + c_j h) + \mathcal{O}(h^{p+1})$$

means that the quadrature order is p. Usually the stage order is of no concern for non-stiff differential equations, and only the quadrature order matters. This is important for explicit methods, since the first step of an explicit method is essentially a step of the explicit Euler method; this means that the stage order for explicit methods is one. Nevertheless, for nonstiff differential equations, we have Runge–Kutta methods of arbitrarily high-order.

On the other hand, stiffness means that the stage order cannot be ignored. Going back to the test equation (9.11), write

$$\Delta_{n,i} = g(t_n + c_i h) - g(t_n) - h \sum_{j=1}^{s} a_{ij} g'(t_n + c_j h),$$

$$\widehat{\Delta}_n = g(t_n + h) - g(t_n) - h \sum_{j=1}^{s} b_j g'(t_n + c_j h).$$

Then, after some calculation, we find that

$$y_{n+1} - g(t_{n+1}) = R(h\lambda) [y_n - g(t_n)] - h\lambda \, \mathbf{b}^T (I - h\lambda A)^{-1} \Delta_n - \widehat{\Delta}_n.$$

Clearly we still need $|R(h\lambda)| \le 1$ for stability. But we have to be careful about Δ_n (the stage errors) as well as $\widehat{\Delta}_n$ (the quadrature error). In other words, our accuracy can be reduced by a low stage order as well as by a low quadrature order.

Many Runge–Kutta methods for stiff differential equations are *stiffly accurate*. This simply means that the last row of A is $\mathbf{b}^T$; that is, $a_{is} = b_i$ for $i = 1, 2, \ldots, s$. An example is the trapezoidal rule:

$$y_{n+1} = y_n + \tfrac{1}{2} h \left[f(t_n, y_n) + f(t_{n+1}, y_{n+1}) \right].$$

The quadrature order is 2 ($\widehat{\Delta}_n = \mathcal{O}(h^3)$), which is the same order as the second stage ($\Delta_{n,2} = \mathcal{O}(h^3)$). The order of the first stage is infinite: $\Delta_{n,1} = 0$, since $c_1 = 0$ and $g(t_n + 0h) = g(t_n) + 0$. For the test equation (9.10), we have

$$y_{n+1} - g(t_{n+1}) = R(h\lambda) [y_n - g(t_n)] - h\lambda \, \mathbf{b}^T (I - h\lambda A)^{-1} \Delta_n - \widehat{\Delta}_n$$

as before. For this method

$$-h\lambda \, \mathbf{b}^T (I - h\lambda A)^{-1} \Delta_n = \frac{2h\lambda}{h\lambda - 2} \left[\frac{1}{2}, \frac{1}{2} \right] \begin{bmatrix} 1 - \frac{1}{2} h\lambda & 0 \\ \frac{1}{2} h\lambda & 1 \end{bmatrix} \begin{bmatrix} 0 \\ \mathcal{O}(h^3) \end{bmatrix}$$

$$= \frac{h\lambda}{h\lambda - 2} \mathcal{O}(h^3).$$

So the *stiff order* of the trapezoidal method is 2, the same as its "normal" order. This is a desirable trait, but it is not shared by most higher-order methods.

Consider, for example, the Gauss methods. The s-stage Gauss method has order $2s$. However, its stiff order is only s. A simple example is the $s = 1$ Gauss method, which is also known as the *midpoint method*, as shown in Table 9.3. Then

$$-h\lambda \, \mathbf{b}^T (I - h\lambda A)^{-1} \Delta_n = -\frac{2h\lambda}{2 - h\lambda} \mathcal{O}(h^2).$$

So while the quadrature order of the midpoint rule is 2, its stiff order is 1. Further analysis for the other Gauss methods can be found in [44].

DIRK methods of any number of stages have stage order ≤ 2, and so the stiff order (for arbitrary $h\lambda$) is ≤ 2. Radau IIA methods with s stages have order $2s - 1$, but the stiff order (for arbitrary $h\lambda$) is $s + 1$. In fact, the global error for Radau IIA methods is $\mathcal{O}(h^{s+1}/(h\lambda))$. If we consider only the case $h\lambda \to \infty$ and $h \to 0$, we find that, because the Radau IIA methods are stiffly accurate, we again get $\mathcal{O}(h^{2s-1})$ global error *in the limit* as $h\lambda \to \infty$. This turns out to be very useful for differential algebraic equations, the topic discussed in Chapter 10. However, for solving problems such as the heat equation (see Section 8.1), there are many eigenvalues λ, some small and some large. So we cannot assume that $h\lambda \to \infty$.

On the other hand, order reduction does not occur for BDF methods. While a complete answer is beyond the scope of this book, consider the differential equation

$$Y' = \lambda \left(Y - g(t) \right) + g'(t) \qquad Y(t_0) = g(t_0).$$

The exact solution is $Y(t) = g(t)$ for all t. If we applied a BDF method to this equation, we get

$$y_{n+1} = \sum_{j=0}^{p-1} a_j y_{n-j} + h\beta \left[\lambda \left(y_{n+1} - g(t_{n+1}) \right) + g'(t_{n+1}) \right].$$

If $e_k = y_k - g(t_k)$ were the error at timestep k, after some algebra we would get

$$(1 - h\lambda) \, e_{n+1} - \sum_{j=0}^{p-1} a_j e_{n-j} = \sum_{j=0}^{p-1} a_j g(t_{n-j}) + h\beta \, g'(t_{n+1}) - g(t_{n+1})$$
$$= \mathcal{O}(h^{p+1}),$$

since the BDF method has order p. But for $h\lambda$ in the stability region, this means that $e_n = \mathcal{O}(h^p)$; if $|h\lambda| \to \infty$ along the negative real axis, then $e_n = \mathcal{O}(h^p/|h\lambda|)$.

9.4 RUNGE–KUTTA METHODS FOR STIFF EQUATIONS IN PRACTICE

While a great many Runge–Kutta methods have been developed, for stiff differential equations, the field narrows to a relatively small number of methods, all of which have the desirable characteristics of stability (especially B-stability) and accuracy (when order reduction is taken into account). The Radau IIA methods score well on just about every characteristic, as they are B-stable, are stiffly accurate and have a high order, even after order reduction is taken into account.

The downside is that Radau methods, like Gauss methods, are expensive to implement. For stiff differential equations, we cannot expect to solve the Runge–Kutta equations by simple iteration. Some sort of nonlinear equation solver is needed. Newton's method is the most common method, but simplified versions of Newton's method are often used in practice, as discussed in Section 8.5 in Chapter 8. For large-scale systems of differential equations, even implementing Newton's method can be difficult as large linear systems need to be solved. This can be done efficiently using the tools of numerical linear algebra. This is an exciting and interesting area in itself, but beyond the scope of this book.

Practical codes for a number of these methods, such as the three-stage, fifth-order Radau IIA method, have been carefully designed, implemented, and tested. An example is the Radau and Radau5 codes of Hairer. For more details see p. 183. These codes are automatic methods that can adjust the stepsize to achieve a user-specified error tolerance.

PROBLEMS

1. Show that the Gauss methods with $s = 1$ and $s = 2$ stages have stiff order s.

2. Consider the following iterative method for solving the Runge–Kutta equations

$$z_{n,i} = y_n + h \sum_{h=1}^{s} a_{ij} f(t_n + c_j h, z_{n,j}), \quad i = 1, 2, \ldots, s.$$

We set

$$z_{n,i}^{(k+1)} = y_n + h \sum_{j=1}^{s} a_{ij} f(t_n + c_j h, z_{n,j}^{(k)}), \quad i = 1, 2, \ldots, s,$$

for $k = 0, 1, 2, \ldots$. Show that if $f(t,x)$ is Lipschitz in x with Lipschitz constant L, then this method is a contractive interation mapping provided

$$hL \max_{1 \leq i \leq s} \sum_{j=1}^{s} |a_{ij}| < 1.$$

Is this method useful for stiff problems?

3. Show that the Gauss methods with $s = 1$ and $s = 2$ are B-stable using the algebraic condition (9.8)–(9.9).

4. Repeat Problem 4 for the Radau IIA methods for $s = 1$ and $s = 2$.

5. Show that the DIRK method in Table 9.8 is *not* B-stable.

6. Show that

$$f(t, y) = \begin{bmatrix} \alpha & +\beta \\ -\beta & \alpha \end{bmatrix} y$$

satisfies a one-sided Lipschitz condition with $\mu \geq \alpha$. Use this to prove that B-stability implies A-stability.

Hint: First show that the eigenvalues of the matrix defining f are $\alpha \pm i\beta$.

7. The one-stage Gauss method is

$$z_{n,1} = y_n + \tfrac{1}{2} h f(t_n + \tfrac{1}{2} h, z_{n,1}),$$
$$y_{n+1} = y_n + h f(t_n + \tfrac{1}{2} h, z_{n,1}).$$

Find the Taylor series expansion of $\widehat{\Delta}_{n,1} = g(t_n + c_1 h) - g(t_n) - h\,a_{11}\,g'(t_n + c_1 h)$ $(c_1 = a_{11} = \frac{1}{2})$ to show that the stage order of this method is 1 while the quadrature order of the method is 2.

8. Derive the coefficients for the Lobatto IIIC method with three stages ($s = 3$, order $= 2s - 2 = 4$). The quadrature points are $c_1 = 0$, $c_2 = \frac{1}{2}$, and $c_3 = 1$. Use the simplifying conditions $B(2s - 2)$ to compute the b_i values, and the simplifying conditions $C(s - 1)$ and one of the conditions in $D(s - 1)$ to compute the a_{ij} matrix entries.

CHAPTER 10

DIFFERENTIAL ALGEBRAIC EQUATIONS

In Chapter 3 we considered the motion of a pendulum consisting of a mass m at the end of a light rigid rod of length l; see Figure 3.1. Deriving the differential equation for the angle θ involved computing the torque about the pivot point. In simple systems like this, it is fairly easy to derive the differential equation from a good knowledge of mechanics. But with more complex systems it can become difficult just to obtain the differential equation to be solved.

Here we will consider a different way of handling this problem that makes it much easier to derive a mathematical model, but at a computational cost. These models contain not only differential equations but also "algebraic" equations. Here "algebraic" does not signify that only the usual operations of arithmetic $(+, -, \times, \text{and} /)$ can appear; rather, it means that no derivatives or integrals of unknown quantities can appear in the equation. Differential and algebraic equations are collectively referred to as *differential algebraic equations* or by the acronym DAE. A number of texts deal specifically with DAEs, such as Ascher and Petzold [10] and Brenan et al. [19].

In this new framework, the position of the mass is given by coordinates (x, y) relative to the pivot for the pendulum. There is a constraint due to the rigid rod: $\sqrt{x^2 + y^2} = l$. There are also two forces acting on the mass. One is gravitation, which acts downward with strength $-mg$. The other is the force that the rod exerts on the mass to maintain the constraint. This force is in the direction of the rod; let its magnitude be N, so that the force itself is $(-Nx, -Ny)/\sqrt{x^2 + y^2}$. This provides a complete model for the pendulum:

$$m\frac{d^2x}{dt^2} = -N\frac{x}{\sqrt{x^2 + y^2}}, \tag{10.1}$$

$$m\frac{d^2y}{dt^2} = -N\frac{y}{\sqrt{x^2 + y^2}} - mg, \tag{10.2}$$

$$0 = l - \sqrt{x^2 + y^2}. \tag{10.3}$$

This second-order system can be rewritten as a first-order system:

$$x' = u, \tag{10.4}$$

$$y' = v, \tag{10.5}$$

$$mu' = -N\frac{x}{\sqrt{x^2 + y^2}}, \tag{10.6}$$

$$mv' = -N\frac{y}{\sqrt{x^2 + y^2}} - mg, \tag{10.7}$$

$$0 = l - \sqrt{x^2 + y^2}. \tag{10.8}$$

The unknowns are the coordinates $x(t)$, $y(t)$, their velocities $u(t)$ and $v(t)$, and the force exerted by the rod is $N(t)$. All in all, there are five equations and five unknown functions. However, only four of the equations are differential equations. The last is an "algebraic" equation. Also, there is no equation with dN/dt in it, so N is called an *algebraic variable*.

For simplicity, we will write $\lambda = N/(m\sqrt{x^2 + y^2})$ so that $du/dt = -\lambda x$ and $dv/dt = -\lambda y - g$. Also, the constraint equation will be replaced by

$$0 = l^2 - x^2 - y^2.$$

We can turn the differential algebraic equations into a pure system of differential equations. To do that, we need to differentiate the algebraic equation until we can obtain an expression for $d\lambda/dt$. Differentiating the constraint three times gives first

$$0 = \frac{d}{dt}\left(l^2 - x^2 - y^2\right) = -2xu - 2yv, \tag{10.9}$$

$$0 = \frac{d^2}{dt^2}\left(l^2 - x^2 - y^2\right) = -2(u^2 + v^2) + 2\lambda(x^2 + y^2) + 2yg, \tag{10.10}$$

and then

$$0 = \frac{d^3}{dt^3}\left(l^2 - x^2 - y^2\right) = 2\frac{d\lambda}{dt}\left(x^2 + y^2\right) + 6gv. \tag{10.11}$$

The number of times that the algebraic equations of a DAE need to be differentiated in order to obtain *differential equations* for all of the algebraic variables is called the *index* of the DAE. Two differentiations allow us to find λ in terms of x, y, u, and v. But three differentiations are needed to compute $d\lambda/dt$ in terms of these quantities. So our pendulum problem is an index 3 DAE.

Solving for λ from the second derivative of the constraint gives

$$\lambda = \frac{u^2 + v^2 - yg}{x^2 + y^2} = \frac{u^2 + v^2 - yg}{l^2}. \tag{10.12}$$

Substituting this expression gives a system of ordinary differential equations:

$$x' = u, \tag{10.13}$$

$$y' = v, \tag{10.14}$$

$$u' = -\frac{u^2 + v^2 - yg}{l^2}x, \tag{10.15}$$

$$v' = -\frac{u^2 + v^2 - yg}{l^2}y - g. \tag{10.16}$$

If, instead of substituting for λ, we differentiate the constraint a third time, we obtain a differential equation for λ:

$$x' = u, \tag{10.17}$$

$$y' = v, \tag{10.18}$$

$$u' = -\lambda x, \tag{10.19}$$

$$v' = -\lambda y - g, \tag{10.20}$$

$$\lambda' = -\frac{3gv}{l^2}. \tag{10.21}$$

The general scheme for a system of differential algebraic equations is

$$Y' = f(t, Y, Z), \qquad Y(t_0) = Y_0, \tag{10.22}$$

$$0 = g(t, Y, Z). \tag{10.23}$$

The Y variables are the differential variables, while the Z variables are the algebraic variables.

10.1 INITIAL CONDITIONS AND DRIFT

In the general scheme, the constraints $0 = g(t, Y, Z)$ must hold at time $t = t_0$, so that $g(t_0, Y_0, Z_0) = 0$, where $Z_0 = Z(t_0)$. So the algebraic variables must also have the right initial values. But the conditions do not stop there. In addition, differentiating the constraints once at $t = t_0$ gives

$$\frac{d}{dt}g(t, Y, Z)|_{t=t_0} = 0,$$

and differentiating twice gives

$$\frac{d^2}{dt^2} g(t, Y, Z)|_{t=t_0} = 0,$$

and so on. This gives a whole sequence of extra initial conditions that must be satisfied. Fortunately, the number of extra conditions is not infinite: the number of differentiatons needed to obtain the needed extra conditions is one less than the *index* of the problem.

Consider, for example, the pendulum problem. Initially the position of the mass is constrained by the length of the rod: $x(t_0)^2 + y(t_0)^2 = l^2$. Differentiating the length constraint (10.8) at $t = t_0$ gives

$$0 = x(t_0)u(t_0) + y(t_0)v(t_0);$$

that is, the initial velocity must be tangent to the circle that the pendulum sweeps out. Finally, the initial force $N(t_0)$ (or equivalently $\lambda(t_0)$) must be set correctly in order for the solution to follow the circle $x^2 + y^2 = l^2$. This gives a total of three extra conditions to satisfy for the initial conditions, coming from the constraint function and its first and second derivatives.

Note that the constraint and the subsequent conditions hold not only at the initial time, but also at any instant. Thus the differential equations obtained that have the algebraic constraint removed (such as (10.13)–(10.16) and (10.17)–(10.20)) must satisfy these additional conditions at all times. Numerical methods do not necessarily preserve these properties even though they are preserved in the differential equations. This is known as *drift*. In theory, if a numerical method for a differential equation or DAE is convergent, then as the stepsize h goes to zero, the amount of drift will also go to zero on any fixed time interval. In practice, however, instabilities that may be introduced by the DAE or ODE formulation mean that extremely small stepsizes may be needed to keep the drift sufficiently small for meaningful answers.

Figure 10.1 shows plots of the trajectories for the pendulum problem using the formulation (10.13)–(10.16) and the Euler and Heun methods (see (4.29)) for its solution.

There are a number of ways of dealing with drift.

1. *Project current solution back to the constraints, either at every step, or occasionally.* For the pendulum example, this means projecting not only the positions (x, y) back to $x^2 + y^2 = l^2$, but also the velocities. Moreover, if λ is computed via a differential equation, it, too, must be projected onto its constraints. Care must be taken in doing this, particularly for multistep methods where projecting just the current solution vector z_n will introduce errors in the approximate solution. Instead, all solution vectors z_{n-j} for $j = 0, 1, \ldots, p$ should be projected, where p is the number of previous iterates used by the multistep method. Also, if the index is high, we should project not only the solution vector, but also the derivative and (if the index is high enough) higher-order derivatives as well onto the appropriate manifold.

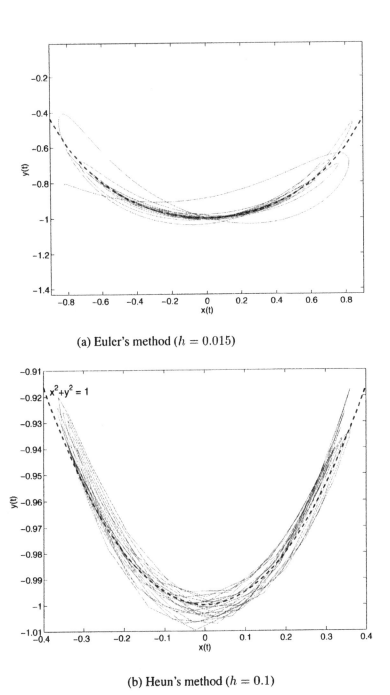

(a) Euler's method ($h = 0.015$)

(b) Heun's method ($h = 0.1$)

Figure 10.1 Plots of trajectories for (10.13)–(10.16) showing drift for Euler and Heun's methods

2. *Modify the differential equation to make the constraint set stable, but otherwise do not change the trajectories.* This technique has been used in a number of contexts, but it almost always has to be done separately for every new case. An example of this technique is the method of Baumgarte [15] for equality-constrained mechanical systems. This would replace the condition $g(t, Y, Z) = 0$ with a differential equation, such as $(d/dt)g(t, Y, Z) + \alpha g(t, Y, Z) = 0$ with $\alpha > 0$; that is

$$0 = g_t(t, Y, Z) + g_y(t, Y, Z) f(t, Y, Z) + g_z(t, Y, Z)Z' + \alpha g(t, Y, Z),$$

which can be solved to give a differential equation for Z. (Note that $g_y(t, Y, Z)$ is the Jacobian matrix of $g(t, Y, Z)$ with respect to Y. See (10.3) below.) For index 3 systems, such as those arising in mechanics, stable second-order equations must be used such as

$$\left(\frac{d^2}{dt^2} + \alpha \frac{d}{dt} + \beta \right) g(t, Y, Z) = 0$$

with suitable choices for α and β. These modifications need to be done with care to ensure that they really are stable, not just for the continuous problem but also for the numerical discretization. Since these stabilization methods have one or more free scaling parameter(s) α (and β), these must be chosen with care. For more information about dealing with these issues, see Ascher et al. [5].

3. *Use a numerical method that explicitly respects the constraints.* These methods treat the differential algebraic equations *as* differential algebraic equations. Instead of necessitating one or more differentiations in order to find differential or other equations for the "algebraic" variables, they are automatically computed by the method itself. These have been developed for general low-index DAEs. Petzold, who developed the first such methods, produced a package DASSL (see [19], [21], [65]) based on backward differentiation formulas (BDFs) for solving index 1 DAEs. Many other methods have been developed, but these tend to be limited in terms of the index that they can handle. All such methods are implicit, and so require the solution of a linear or nonlinear system of equations at each step.

To summarize: methods 1 and 2 for handling DAEs have some problems. The projection method can work with some ODE methods. The Baumgarte stabilization method can also be made to work, but requires "tuning" the stabilization parameters; this method can run into trouble for stiff equations. Method 3, designing numerical methods that explicitly recognize the constraints, is the one that we focus on in the remainder of the chapter.

10.2 DAES AS STIFF DIFFERENTIAL EQUATIONS

Differential algebraic equations can be treated as the limit of ordinary differential equations. Note that $g(t, Y, Z) = 0$ if and only if $Bg(t, Y, Z) = 0$ for any nonsingular square matrix B. Then the DAE (10.22)–(10.23) can be treated as the limit as $\epsilon \to 0$ of

$$Y' = f(t, Y, Z), \qquad Y(t_0) = Y_0, \tag{10.24}$$
$$\epsilon Z' = B(Y)g(t, Y, Z). \tag{10.25}$$

The matrix function $B(Y)$ should be chosen to make the differential equation in Z (10.25) stable, so that the solution for (10.25), $Z(t)$, converges to the solution $Z = Z^*$ where $g(t, Y, Z^*) = 0$.

For ϵ small, these equations are *stiff*, so implicit methods are needed. Furthermore, since the order obtained in practice for an implicit method can differ from the order of the method for nonstiff problems, the order of an implicit method may deviate from the usual order when it is applied to differential algebraic equations.

But how do we apply a numerical method for stiff ODEs to a DAE? The simplest method to apply is the implicit Euler method. If we apply it to the stiff approximation (10.24)–(10.25) using step size h, we get

$$y_{n+1} = y_n + h\,f(t_{n+1}, y_{n+1}, y_{n+1}), \tag{10.26}$$
$$\epsilon z_{n+1} = \epsilon z_n + h\,B(y_{n+1})g(t_{n+1}, y_{n+1}, z_{n+1}). \tag{10.27}$$

Taking the limit as $\epsilon \to 0$ and recalling that $B(Y)$ is nonsingular, we get the equations

$$y_{n+1} = y_n + h\,f(t_{n+1}, y_{n+1}, z_{n+1}), \tag{10.28}$$
$$0 = g(t_{n+1}, y_{n+1}, z_{n+1}). \tag{10.29}$$

This method will work for index 1 DAEs, but not in general for higher index DAEs.

An issue regarding accuracy is the *stiff order* of an ODE solver: the order of a method for solving stiff ODEs may be lower than that for solving a nonstiff ODE, as noted in Section 9.3. Since DAEs can be considered to be an extreme form of stiff ODEs, this can also affect DAE solvers. With some methods, some components of the solution (e.g., positions) can be computed more accurately than other components (e.g., forces).

10.3 NUMERICAL ISSUES: HIGHER INDEX PROBLEMS

Consider index 1 problems in standard form:

$$Y' = f(t, Y, Z), \qquad Y(t_0) = Y_0,$$
$$0 = g(t, Y, Z).$$

Here $Y(t)$ is an n-dimensional vector and $Z(t)$ is an m-dimensional vector. The function

$$g(t, Y, Z) = [g_1(t, T, Z), g_2(t, Y, Z), \ldots, g_m(t, Y, Z)]^T$$

must have values that are m-dimensional vectors. For an index 1 problem, the Jacobian matrix of $g(t, Y, Z)$ with respect to Z, specifically

$$g_z(t, Y, Z) = \begin{bmatrix} \partial g_1/\partial z_1 & \partial g_1/\partial z_2 & \cdots & \partial g_1/\partial z_m \\ \partial g_2/\partial z_1 & \partial g_2/\partial z_2 & \cdots & \partial g_2/\partial z_m \\ \vdots & \vdots & \ddots & \vdots \\ \partial g_m/\partial z_1 & \partial g_m/\partial z_2 & \cdots & \partial g_m/\partial z_m \end{bmatrix}_{(t,Y,Z)}$$

is nonsingular. So we can apply the implicit function theorem to show that whenever $g(t_0, y_0, z_0) = 0$, there is locally a smooth solution function $z = \varphi(t, y)$, where $z_0 = \varphi(t_0, y_0)$. With a numerical solution (y_n, z_n), $n = 0, 1, 2, \ldots$, the error in z_n should be of the same order as the error in y_n. This does not always happen, but requires some special properties of the numerical method. As we will see for Runge–Kutta methods, we need the method to be stiffly accurate. A method is stiffly accurate when the last row of the A matrix in the Butcher tableau is the same as the bottom row b^T of the Butcher tableau. Stiff accuracy is important for understanding Runge–Kutta methods for stiff differential equations, as was noted in Section 9.3.

Index 2 problems have a standard form:

$$Y' = f(t, Y, Z), \qquad Y(t_0) = Y_0, \tag{10.30}$$
$$0 = g(t, Y), \tag{10.31}$$

where the product of Jacobian matrices of $g_y(t, Y) f_z(t, Y, Z)$ is nonsingular. But now, to determine $Z(t)$, we need dY/dt. Thus numerical methods applied to index 2 problems will need to perform some kind of "numerical differentiation" in order to find $Z(t)$. This may result in a reduction of the order of accuracy in the numerical approximation $Z(t)$, which can feed back into the equation (10.30) for $Y(t)$.

Index 3 problems, such as our pendulum problem, require more specialized treatment. These problems are discussed in Subsection 10.6.1. However, the same complication arises — different components of the solution can have different orders of convergence.

To illustrate this complication, consider the problem of the spherical pendulum. This is just like the ordinary planar pendulum, except that the mass is not constrained to a single vertical plane. This is sometimes called "Foucault's pendulum", and can be used to demonstrate the rotation of the earth, although our model will not include that effect. For this system, we use $q = [x, y, z]^T$ for the position of the mass m, which is subject to the constraint that $q^T q = \ell^2$ and a downward gravitational force of strength mg. Using the methods of Subsection 10.6.1, we obtain the following index 3 DAE:

$$mv' = -\lambda q - m g k,$$
$$q' = v,$$
$$0 = \frac{1}{2}(q^T q - \ell^2),$$

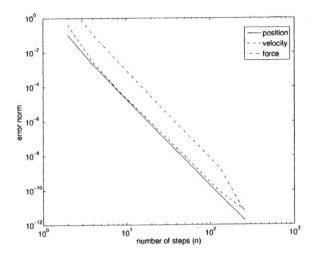

Figure 10.2 Errors in solving the spherical pendulum problem using the three-stage Radau IIA method with an index 1 DAE

where **k** is the unit vector pointing upward. Note that the state vector for the DAE is $\mathbf{y}^T = [\mathbf{q}^T, \mathbf{v}^T, \lambda]$.

By differentiating the constraints as we did for the planar pendulum, we can obtain lower index DAEs. If we differentiate the constraint once, we obtain

$$0 = \mathbf{v}^T \mathbf{q}$$

to give an index 2 DAE. If we differentiate again, we obtain

$$0 = \mathbf{v}^T \mathbf{v} - \frac{\lambda}{m} \mathbf{q}^T \mathbf{q} - \mathbf{k}^T \mathbf{q} g$$

to give an index 1 DAE.

Using the Radau IIA method with three stages (which is normally fifth-order), we can solve each of these systems. Figures 10.2–10.4 show the numerical results for each of these DAEs with indices 1, 2 and 3. The specific parameter values used are $m = 2$ and $\ell = \frac{3}{2}$; the initial time was $t = 0$, and the errors were computed at $t = 1$. As can be clearly seen, for both index 2 and index 3 cases, the forces are computed considerably less accurately than are the other components, and the slope of the error line for the forces (λ) is substantially less than those for the other components. This indicates a lower-order of convergence for the forces in the index 2 and index 3 versions of the problem. For the index 3 case, both the forces and velocities (**v**) appear to have a lower-order of convergence than the positions (**q**). However, the order of convergence of the positions does not seem to be affected by the index of the DAE.

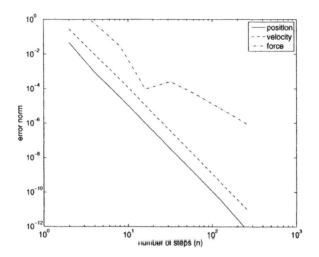

Figure 10.3   Errors in solving the spherical pendulum problem using the three-stage Radau IIA method with an index 2 DAE

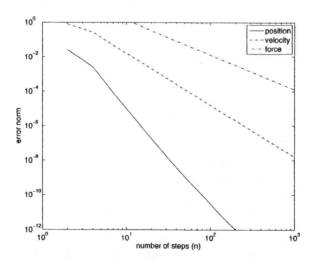

Figure 10.4   Errors in solving the spherical pendulum problem using the three-stage Radau IIA method with an index 3 DAE

From these numerical results, the following question may arise: Why use high index DAEs? As noted above, one reason is that using the high index formulation can prevent drift in the main constraint $g(\mathbf{q}) = 0$. Another reason is that the model of the spherical pendulum is most naturally given as an index 3 DAE. The lower index DAEs are constructed by differentiating the constraint function. While this is often the quickest approach for simple problems, for large problems this can become difficult to do, and might not be possible in practice for functions defined by some (complicated) piece of code.

10.4 BACKWARD DIFFERENTIATION METHODS FOR DAES

The first ODE methods to be applied to DAEs were the backward differentiation formula (BDF) methods. These work well for index 1 DAEs, and are the basis of the code DASSL [19], [65]. These implicit methods were introduced in Section 8.2 and have the form

$$y_{n+1} = \sum_{j=0}^{p-1} a_{n-j}\, y_{n-j} + h\beta\, f(t_{n+1}, y_{n+1}).$$

The coefficients a_j and β are chosen so that

$$y'(t_{n+1}) = \frac{1}{\beta h}\left[y_{n+1} - \sum_{j=0}^{p-1} a_j y_{n-j} \right] + \mathcal{O}(h^p),$$

giving a method of order p.

These methods, while not A-stable, are nevertheless very well behaved, at least for nonoscillatory problems for $p \leq 6$. If $p \geq 7$, part of the negative real axis lies outside the stability region, and the method can become unstable for $\lambda < 0$ large enough to put $h\lambda$ in the unstable region. For this reason, we restrict $p \leq 6$ for BDF methods.

10.4.1 Index 1 problems

For DAEs of the form

$$Y' = f(Y, Z), \qquad Y(t_0) = Y_0, \tag{10.32}$$
$$0 = g(Y, Z), \tag{10.33}$$

the BDF method becomes

$$y_{n+1} = \sum_{j=0}^{p-1} a_j y_{n-j} + h\beta\, f(y_{n+1}, z_{n+1}),$$
$$0 = g(y_{n+1}, z_{n+1}).$$

For index 1 DAEs, the equation $g(y, z) = 0$ gives z implicitly as a function of y. If we write $z = \varphi(y)$ as this implicit function, the BDF method can be reduced to

$$y_{n+1} = \sum_{j=0}^{p} a_j y_{n-j} + h\beta\, f(y_{n+1}, \varphi(y_{n+1})),$$

which is the result of applying the BDF method to the reduced equation

$$Y' = f(y, \varphi(Y)).$$

Thus the BDF method gives a numerical solution with the expected rate of convergence to the true solution.

10.4.2 Index 2 problems

BDF methods can be used for DAEs of index 2 as well as index 1, particularly for the semi-explicit index 2 DAEs:

$$Y' = f(Y, Z), \qquad Y(t_0) = y_0, \tag{10.34}$$
$$0 = g(Y). \tag{10.35}$$

Recall that $g(Y)$ is an m-dimensional vector for each Y, so that $g_y(Y)$ is an $m \times n$ matrix. On the other hand, $f(Y, Z)$ is an n-dimensional vector, so that $f_z(Y, Z)$ is an $n \times m$ matrix. The product $g_y(Y)\, f_z(Y, Z)$ is thus an $m \times m$ matrix. We assume that $g_y(Y)\, f_z(Y, Z)$ is nonsingular.

The DAE (10.34)–(10.35) is index 2 if we can (locally) solve for $Z(t)$ from $Y(t)$ using only one differentiation of the "algebraic" equation $g(Y) = 0$. Differentiating gives $0 = g_y(Y)\, dY/dt = g_y(Y)\, f(Y, Z)$. So for an index 2 DAE, the function $Z \mapsto g_y(Y)\, f(Y, Z)$ needs to be invertible so that we can find a smooth implicit function $Y \mapsto Z$. The usual requirement needed is that the Jacobian matrix of the map $Z \mapsto g_y(Y)\, f(Y, Z)$ be an invertible matrix on the exact solution. From the usual rules of calculus, this comes down to requiring that $g_y(Y(t))\, f_z(Y(t), Z(t))$ is an invertible matrix for all t on the exact solution. Note that this implies that $g_y(Y)\, f_z(Y, Z)$ is invertible for any (Y, Z) *sufficiently near* the exact solution as well.

Assuming that $g_y(Y)\, f_z(Y, Z)$ is nonsingular, we can show that the p-step BDF method for DAEs,

$$y_{n+1} = \sum_{j=0}^{p-1} \alpha_j\, y_{n-j} + h\beta\, f(y_{n+1}, z_{n+1}),$$
$$0 = g(y_{n+1}),$$

is convergent of order p for $p \le 6$. Recall that for $p \ge 7$, the stability region for the p-step BDF method *does not* include all of the negative real axis, making it unsuitable for stiff ODEs or DAEs.

It should be noted that these methods are implicit, and therefore require the solution of a nonlinear system of equations. We can use Newton's method or any number of variants thereof [55]. The system of equations to be solved has $n + m$ equations and $n + m$ unknowns.

For the p-step BDF method, we have

$$y_n - Y(t_n) = \mathcal{O}(h^p),$$
$$z_n - Z(t_n) = \mathcal{O}(h^p),$$

provided $y_j - Y(t_j) = \mathcal{O}(h^{p+1})$ for $j = 0, 1, 2, \ldots, p - 1$ ([20], [40], [44], [60]). Note that we need one order higher accuracy in the *initial* values; this is necessary as our estimates for $z_j, j = 0, 1, \ldots, p - 1$, are essentially obtained by differentiating the data for $y_j, j = 0, 1, \ldots, p - 1$.

Note that it is particularly important to solve the equations $g(y_{n+1}) = 0$ accurately. Noise in the solution of these equations will be amplified by a factor of order $1/h$ to produce errors in z_{n+1}. This, in turn, will result in larger errors in y_n over time.

10.5 RUNGE–KUTTA METHODS FOR DAES

As for stiff equations, the Runge–Kutta methods used for DAEs need to be implicit methods. The way that a Runge–Kutta method is used for the index 1 DAE (10.32)–(10.33),

$$Y' = f(Y, Z), \qquad Y(t_0) = Y_0, \tag{10.36}$$
$$0 = g(Y, Z), \tag{10.37}$$

is

$$y_{n,i} = y_n + h \sum_{j=1}^{s} a_{ij} f(y_{n,j}, z_{n,j}), \tag{10.38}$$

$$0 = \sum_{j=1}^{s} a_{ij} g(y_{n,j}, z_{n,j}), \tag{10.39}$$

$$y_{n+1} = y_n + h \sum_{j=1}^{s} b_j f(y_{n,j}, z_{n,j}), \tag{10.40}$$

for $i = 1, 2, \ldots, s$. Provided the matrix A is invertible, (10.39) is equivalent to

$$0 = g(y_{n,i}, z_{n,i}), \qquad i = 1, 2, \ldots, s.$$

As for BDF methods, these are systems of nonlinear equations, and can be solved by Newton's method or its variants [55]. Unlike the BDF methods, the number of equations to be solved are $s(M + N)$ with $s(M + N)$ unknowns where Y is a vector with N components and Z has M unknowns.

Also, the analysis of error in stiff problems in Section 9.3 shows that the stage order of the Runge–Kutta method essentially determines the order of the Runge–Kutta method for DAEs. For this to work well, we usually require that the method be *stiffly accurate* (such as Radau IIA methods); that is, $\mathbf{b}^T$ must be the bottom row of A: $b_i = a_{s,i}$ for $i = 1, 2, \ldots, s$. This means that $y_{n+1} = y_{n,s}$ and setting $z_{n+1} = z_{n,s}$ so that $g(y_{n+1}, z_{n+1}) = 0$. As with stiff equations, the stability function $R(h\lambda) = 1 + h\lambda \, \mathbf{b}^T \left(I - h\lambda \, A\right)^{-1} \mathbf{e}$ (see (9.7)) gives crucial information about the behavior of the method. However, for DAEs, we are considering what happens as $h\lambda \rightarrow -\infty$. Since $R(h\lambda)$ is a rational function of $h\lambda$, the important quantity is $R(\infty) = R(-\infty) = 1 - \mathbf{b}^T A^{-1} \mathbf{e}$ for nonsingular A.

10.5.1   Index 1 problems

Consider index 1 problems of the form

$$Y' = f(Y, Z), \quad Y(t_0) = Y_0,$$
$$0 = g(Y, Z).$$

Let us suppose that we have an implicit function φ for g, meaning that whenever $0 = g(y, z)$, then $z = \varphi(y)$. If we can do this, then the problem reduces to finding the solution of

$$Y' = f(Y, \varphi(Y)), \quad Y(t_0) = Y_0.$$

Note that if the Jacobian matrix $\nabla_z f(y^*, z^*)$ is nonsingular, then we can find a *local* implicit function φ so that $\varphi(y^*) = z^*$ and φ is smooth nearby to y^*. Then in this case, $g(y_{n,i}, z_{n,i}) = 0$ implies that $z_{n,i} = \varphi(y_{n,i})$, and our Runge–Kutta equations imply that

$$y_{n,i} = y_n + h \sum_{j=1}^{s} a_{ij} \, f(y_{n,j}, z_{n,j})$$

$$= y_n + h \sum_{j=1}^{s} a_{ij} \, f(y_{n,j}, \varphi(y_{n,j})).$$

For a stiffly accurate method, $y_{n+1} = y_{n,s}$ and $z_{n+1} = z_{n,s} = \varphi(y_{n,s}) = \varphi(y_{n+1})$. This is exactly what the Runge–Kutta method would give when applied to the ordinary differential equation

$$Y' = f(Y, \varphi(Y)), \quad Y(t_0) = Y_0.$$

So the order of accuracy is exactly what we would expect for smooth ordinary differential equations.

The case where the method is *not* stiffly accurate is a little more complex; the argument for the accuracy of $y_n \approx Y(t_n)$ is not changed, but the accuracy of the computed values $z_n \approx Z(t_n)$ is, and can depend on the value of $R(\infty)$. Recall that p is the quadrature order of the method, and q is the stage order. In terms of the simplifying conditions (9.4)–(9.6), conditions $B(p)$ and $C(q)$ hold. The error

$z_n - Z(t_n) = \mathcal{O}(h^r)$, where $r = \min(p, q+1)$ if $-1 \leq R(\infty) < 1$ and $r = \min(p-1, q)$ if $R(\infty) = 1$; but $z_n - z(t_n)$ diverges exponentially in n if $|R(\infty)| > 1$. We show this below.

Suppose our Runge–Kutta method has stage order q and quadrature order p, so that for a smooth function $\psi(\cdot)$, we obtain

$$\psi(t_n + c_i h) = \psi(t_n) + h \sum_{j=1}^{s} a_{ij} \psi'(t_n + c_j h) + \mathcal{O}(h^{q+1}),$$

$$i = 1, \ldots, s, \qquad (10.41)$$

$$\psi(t_{n+1}) = \psi(t_n) + h \sum_{i=1}^{s} b_i \psi'(t_n + c_j h) + \mathcal{O}(h^{p+1}). \qquad (10.42)$$

The global order of this method for DAEs can be determined from the stage and quadrature orders depending on several cases: (1) the method is stiffly accurate, (2) $-1 \leq R(\infty) < 1$, (3) $R(\infty) = 1$, or (4) $|R(\infty)| > 1$.

If the method is stiffly accurate, then (as we have seen) the accuracy for index 1 DAEs is the same as for smooth ordinary differential equations: $Y(t_n) - y_n = \mathcal{O}(h^p)$, provided $t_n - t_0$ is bounded.

If the method is not stiffly accurate, then the stage order q becomes important. If we write

$$\Psi_n = [\psi(t_n + c_1 h), \psi(t_n + c_2 h), \ldots, \psi(t_n + c_s h)]^T,$$
$$\Psi'_n = [\psi'(t_n + c_1 h), \psi'(t_n + c_2 h), \ldots, \psi'(t_n + c_s h)]^T,$$

then, from (10.41), we obtain

$$\Psi_n = \psi(t_n) \mathbf{e} + h A \Psi'_n + \mathcal{O}(h^{q+1}),$$

so that for nonsingular A, we have

$$\Psi'_n = h^{-1} A^{-1} (\Psi_n - \mathbf{e}\, \psi(t_n)) + \mathcal{O}(h^q).$$

Substituting this into (10.42) gives

$$\psi(t_{n+1}) = \left(1 - \mathbf{b}^T A^{-1} \mathbf{e}\right) \psi(t_n) + \mathbf{b}^T A^{-1} \Psi_n + \mathcal{O}(h^{q+1}) + \mathcal{O}(h^{p+1}).$$

But $1 - \mathbf{b}^T A^{-1} \mathbf{e} = R(\infty)$. Thus

$$\psi(t_{n+1}) = R(\infty)\, \psi(t_n) + \mathbf{b}^T A^{-1} \Psi_n + \mathcal{O}(h^{q+1}) + \mathcal{O}(h^{p+1}).$$

In particular, we can take $\psi(t) = Z(t)$ and $\psi(t) = Y(t)$, giving

$$Z(t_{n+1}) = R(\infty)\, Z(t_n) + \mathbf{b}^T A^{-1} \mathbf{Z}_n + \mathcal{O}(h^{q+1}) + \mathcal{O}(h^{p+1}), \qquad (10.43)$$
$$Y(t_{n+1}) = R(\infty)\, Y(t_n) + \mathbf{b}^T A^{-1} \mathbf{Y}_n + \mathcal{O}(h^{q+1}) + \mathcal{O}(h^{p+1}),$$

with

$$\mathbf{Z}_n = [Z(t_n + c_1 h), \dots, Z(t_n + c_s h)]^T,$$
$$\mathbf{Y}_n = [Y(t_n + c_1 h), \dots, Y(t_n + c_s h)]^T.$$

Now $g(y_{n,i}, z_{n,i}) = 0$ so $z_{n,i} = \varphi(y_{n,i})$ as noted above. Let

$$\widehat{\mathbf{Y}}_n = [y_{n,1}, y_{n,2}, \dots, y_{n,s}]^T,$$
$$\widehat{\mathbf{Z}}_n = [z_{n,1}, z_{n,2}, \dots, z_{n,s}]^T.$$

Then the Runge–Kutta equations can be written (as we did with $\psi(t)$ above) as

$$z_{n+1} = R(\infty) z_n + \mathbf{b}^T A^{-1} \widehat{\mathbf{Z}}_n. \tag{10.44}$$

The error $\Delta z_{n+1} = Z(t_{n+1}) - z_{n+1}$ is given by subtracting the above equations (10.43) and (10.44), yielding

$$\Delta z_{n+1} = R(\infty) \Delta z_n + \mathbf{b}^T A^{-1} \left(\mathbf{Z}_n - \widehat{\mathbf{Z}}_n \right) + \mathcal{O}(h^{q+1}) + \mathcal{O}(h^{p+1}).$$

Note that $z_{n,i} = \varphi(y_{n,i})$ and $Z(t_n + c_i h) = \varphi(Y(t_n + c_i h)$. The stage order is q, so from the differential equation for Y and the Runge–Kutta method,

$$y_{n,i} - Y(t_n + c_i h)$$
$$= y_n - Y(t_n)$$
$$+ h \sum_{j=1}^{s} a_{ij} \left(f(y_{n,j}, \varphi(y_{n,j})) - f(Y(t_n + c_j h), \varphi(Y(t_n + c_j h))) \right) + \mathcal{O}(h^{q+1}).$$

Since $y_n = Y(t_n) + \mathcal{O}(h^p)$, we get

$$y_{n,i} = Y(t_n + c_i h) + \mathcal{O}(h^{\min(p,q+1)}).$$

So

$$z_{n,i} - Z(t_n + c_i h) = \varphi(y_{n,i}) - \varphi(Y(t_n + c_i h)) = \mathcal{O}(h^{\min(p,q+1)}).$$

Therefore

$$\Delta z_{n+1} = R(\infty) \Delta z_n + \mathcal{O}(h^{\min(p,q+1)}).$$

If $|R(\infty)| < 1$, then we obtain the expected global order of z_n. If $R(\infty) = 1$ we the errors can accumulate giving a convergence order of one less. If $|R(\infty)| > 1$, then z_n will grow exponentially in n. If $R(\infty) = -1$, then we need to do some more analysis to show that the hidden constant in the "$\mathcal{O}(h^{\min(p,q+1)})$" is actually a smooth function of t. Then successive steps will cause cancellation of the error, and the global error for z_n is $\mathcal{O}(h^{\min(p,q+1)})$.

To illustrate these theoretical results, consider again the numerical results shown in Figure 10.2 for the index 1 version of the spherical pendulum problem using the 3-stage 5th-order Radau IIA method. All components of the solution converge with roughly the same order of accuracy. In fact, the slopes of the straightest parts of the the graphs in Figure 10.2 are $\approx -5.10, -5.04$, and -5.05 for the position, velocity, and force components of the solution, respectively. This indicates that the index 1 DAE is being solved with the full order of accuracy that the three-stage Radau IIA method can provide.

10.5.2 Index 2 problems

Here we consider index 2 problems of the form

$$Y' = f(Y, Z),$$
$$0 = g(Y).$$

As in Subsection 10.4.2, we assume that $g_y(Y) f_z(Y, Z)$ is a square nonsingular matrix on the exact solution.

Index 2 problems are considerably harder to solve numerically than corresponding index 1 problems. In the index 1 case where the "algebraic" equations $g(Y, Z) = 0$ give Z as a function of Y ($Z = \varphi(Y)$), the result of solving this system of equations could be substituted into $dY/dt = f(Y, Z) = f(Y, \varphi(Y))$ to form a smooth ordinary differential equation. This is not possible in the index 2 case. Indeed, the task of determining whether initial values (y_0, z_0) are consistent (i.e. $g_y(y_0) f(y_0, z_0) = 0$) is a non-trivial task.

Runge–Kutta methods for index 2 problems have the form

$$y_{n,i} = y_n + h \sum_{j=1}^{s} a_{ij} f(y_{n,j}, z_{n,j}), \qquad \text{for } i = 1, 2, \ldots, s,$$

$$z_{n,i} = z_n + h \sum_{j=1}^{s} a_{ij} \ell_{n,j}, \qquad \text{for } i = 1, 2, \ldots, s,$$

$$y_{n+1} = y_n + h \sum_{j=1}^{s} b_j f(y_{n,j}, z_{n,j}),$$

$$z_{n+1} = z_n + h \sum_{j=1}^{s} b_j \ell_{n,j},$$

$$0 = g(y_{n,i}), \qquad \text{for } i = 1, 2, \ldots, s.$$

Note that we have extra variables $\ell_{n,i}$ that are needed to solve the equations $g(y_{n,i}) = 0$. If (y_n, z_n) is sufficiently close to being consistent, there exists (y_{n+1}, z_{n+1}) (as well as the $y_{n,j}, z_{n,j}$, and $\ell_{n,j}$) satisfying the Runge–Kutta equations, and (y_{n+1}, z_{n+1}) is also close to being consistent.

This non-linear system of equations can be solved using, for example, Newton's method. Given currently computed values $y_{n,j}^{(k)}, z_{n,j}^{(k)}, \ell_{n,j}^{(k)}$ and y_n, z_n from the previous step, we compute corrected values $y_{n,j}^{(k+1)} = y_{n,j}^{(k)} + \Delta y_{n,j}, z_{n,j}^{(k+1)} = z_{n,j}^{(k)} + \Delta z_{n,j}$,

Table 10.1 Order of accuracy for index 2 DAEs of the form (10.34)–(10.35) for methods with s stages

Method	y	z
Gauss	$\begin{cases} s+1, & s \text{ odd} \\ s, & s \text{ even} \end{cases}$	$\begin{cases} s-1, & s \text{ odd} \\ s-2, & s \text{ even} \end{cases}$
Radau IIA	$2s-1$	s
Lobatto IIIC	$2s-2$	$s-1$
DIRK a	2	1

and $\ell_{n,j}^{(k+1)} = \ell_{n,j}^{(k)} + \Delta\ell_{n,j}$ by solving the linear system

$$y_{n,i}^{(k)} + \Delta y_{n,i} = y_n + h \sum_{j=1}^{s} a_{ij} \left[f(y_{n,j}^{(k)}, z_{n,j}^{(k)}) + f_y(y_{n,j}^{(k)}, z_{n,j}^{(k)})\Delta y_{n,j} \right.$$
$$\left. + f_z(y_{n,j}^{(k)}, z_{n,j}^{(k)})\Delta z_{n,j} \right], \quad \text{for } i = 1, 2, \ldots, s,$$

$$z_{n,i}^{(k)} + \Delta z_{n,i} = z_n + h \sum_{j=1}^{s} a_{ij} \left[\ell_{n,j}^{(k)} + \Delta\ell_{n,j} \right], \quad \text{for } i = 1, 2, \ldots, s,$$

$$0 = g(y_{n,i}^{(k)}) + g_y(y_{n,i}^{(k)}) \Delta y_{n,i}, \quad \text{for } i = 1, 2, \ldots, s.$$

There are several implications of the theory of these problems for numerical methods, such as Runge–Kutta methods, for index 2 DAEs.

1. The order of accuracy for the numerical solutions $z_n \approx Z(t_n)$ and $y_n \approx Y(t_n)$ are often different.

2. The non-linear systems are generally harder to solve for index 2 systems than for index 1 systems. More specifically, the condition number of the linear system for Newton's method increases as $\mathcal{O}(1/h)$ as the step size h becomes small [44, § VII.4]. By comparison, the linear systems for Newton's method for index 1 DAEs have bounded condition numbers as h goes to zero.

3. Additional conditions are needed to obtain convergence of the numerical methods.

Development of the theory for the order of convergence of these methods is beyond the scope of this book. However, we can present results for some families of Runge–Kutta methods, which are summarized in Table 10.1 ([42]). In the table, the DIRK method is taken from Table 9.8 (a) in Chapter 9 with $s = 3$.

Note that the Gauss methods suffer a strong loss of accuracy, obtaining only order $s + 1$ at best for y (compared to $2s - 1$ for ordinary differential equations), while Radau IIA methods keep the same order for y as for solving ordinary differential equations. The order for z is less for all methods listed, often quite substantially less. One reason for the good performance of Radau IIA methods is that it is stiffly accurate, and has a high stage order (q) as well as having a good quadrature order (p). The Lobatto IIIC method, which is stiffly accurate, also has a good order of accuracy.

One of the most popular methods for solving DAEs is the 5th-order, 3-stage Radau IIA method (Table 9.7). This is the basis for some popular software for DAEs. For more information, see p. 183. Numerical results for this method (with a fixed stepsize) are shown in Figure 10.3 for the index 2 version of the spherical pendulum problem. The slopes of the graphs are $\approx -5.01, -4.98$, and -2.85 for the position, velocity, and force components, respectively. In this version, the force component plays the role of Z, while the position and velocity components play the role of Y. These results seem roughly consistent with the expected fifth-order convergence of y_n to $Y(t)$, and third-order convergence of z_n to $Z(t)$.

Some other Runge–Kutta-type methods have been developed for index 2 DAEs, such as that proposed by Jay [51], which uses separate methods for the Y and Z components of the solution.

10.6 INDEX THREE PROBLEMS FROM MECHANICS

Mechanics is a rich source of DAEs; the pendulum example of Figure 3.1 and (10.1)–(10.3) is a common example. For general mechanical systems, we need a more systematic way of deriving the equations of motion. There are two main ways of doing this: Lagrangian mechanics and Hamiltonian mechanics. Although closely related, they each have their own specific character. We will use the Lagrangian approach here.

For more information about this area, which is often called *analytical mechanics*, see Fowles [38] for a traditional introduction, and Arnold [4] or Marsden and Ratiu [61] for more mathematical treatments. A comprehensive approach can be found in Fasano and Marmi [37], which includes extensions to statistical mechanics and continuum mechanics as well as more traditional topics.

In the Lagrangian approach to mechanics, the main variables are the *generalized coordinates* $q = [q_1, q_2, \ldots, q_n]^T$ and the *generalized velocities* $v = dq/dt$. Note that in this section q is *not* the stage order. The generalized coordinates can be any convenient system of coordinates for representing the configuration of the system. For example, for a pendulum in the plane, we could use either the angle to the vertical θ, or x and y coordinates for the center of mass. In the latter case we will need to include one (or more) constraints on the coordinates: $g(q) = 0$. Note that since the generalized coordinates could include angles, the generalized velocity vector could include angular velocities as well as ordinary velocities.

The function that defines the motion in Lagrangian mechanics is the Lagrangian function $L(q, v)$, a scalar function of the generalized coordinates and generalized velocities. For a system with no constraints on the coordinates, we have

$$L(q, v) = T(q, v) - V(q),$$

where $T(q, v)$ is the kinetic energy of the system and $V(q)$ is the potential energy of the system. Usually the kinetic energy is quadratic in the velocity:

$$T(q, v) = \tfrac{1}{2}v^T M(q)\, v.$$

Here $M(q)$ is the *mass matrix*, although since v may contain quantities such as angular as well as ordinary velocities, the entries in $M(q)$ may include quantities such as moments of inertia as well as ordinary masses. If we have constraints on the coordinates[1], $g(q) = 0$, then these constraints can be incorporated into the Lagrangian function using Lagrange multipliers:

$$L(q, v, \lambda) = T(q, v) - V(q) - \lambda^T g(q).$$

The Lagrange multipliers can be regarded as generalized forces that ensure that the constraints are satisfied. The equations of motion are obtained by means of the Euler–Lagrange equations

$$0 = \frac{d}{dt} L_v(q, v) - L_q(q, v),$$

where $L_v(q, v)$ is the gradient vector of $L(q, v)$ with respect to v, and $L_q(q, v)$ is the gradient vector of $L(q, v)$ with respect to q. If we have constraints $g(q) = 0$, the Euler–Lagrange equations become

$$0 = \frac{d}{dt} L_v(q, v, \lambda) - L_q(q, v, \lambda), \tag{10.45}$$

$$0 = g(q) = L_\lambda(q, v, \lambda). \tag{10.46}$$

For the pendulum example, let us use $q = [x, y]^T$ as the position of the mass, and $v = dq/dt = [dx/dt, dy/dt]^T$ is its velocity. The constraint is

$$g(q) = \frac{1}{2}\left(x^2 + y^2 - \ell^2\right) = 0.$$

The kinetic energy is just the energy of a mass moving with velocity v:

$$T(q, v) = \frac{1}{2}m\left[\left(\frac{dx}{dt}\right)^2 + \left(\frac{dy}{dt}\right)^2\right].$$

The potential energy is just the potential energy due to gravity: $V(q) = mgy$. The Lagrangian is then

$$L(q, dq/dt, \lambda) = \frac{m}{2}\left(\left(\frac{dx}{dt}\right)^2 + \left(\frac{dy}{dt}\right)^2\right) - mgy - \lambda\frac{1}{2}\left(x^2 + y^2 - \ell^2\right).$$

---

[1]Here we have constraints on the generalized coordinates *alone*: $g(q) = 0$. These are called *holonomic* constraints.

The Euler–Lagrange equations are then

$$
0 = \frac{d}{dt}\left(m \begin{bmatrix} \frac{dx}{dt} \\ \frac{dy}{dt} \end{bmatrix} \right) + \begin{bmatrix} 0 \\ mg \end{bmatrix} + \lambda \begin{bmatrix} x \\ y \end{bmatrix},
$$

$$
0 = \frac{1}{2}\left(x^2 + y^2 - \ell^2 \right).
$$

This is essentially the pendulum DAE (10.1)–(10.3) rearranged.

Not only does this DAE have index 3, but all problems of this type have index 3 (or higher). In general, for mechanical systems, the Euler–Lagrange equations become

$$
M(q)\frac{dv}{dt} = k(q, v) - \nabla V(q) - \nabla g(q)^T \lambda, \tag{10.47}
$$

$$
\frac{dq}{dt} = v, \tag{10.48}
$$

$$
0 = g(q), \tag{10.49}
$$

where

$$
k_i(q, v) = \frac{1}{2}\sum_{j,k=1}^{n}\left(\frac{\partial m_{jk}}{\partial q_i} - \frac{\partial m_{ij}}{\partial q_k} - \frac{\partial m_{ik}}{\partial q_j} \right) v_j\, v_k, \qquad i = 1, 2, \ldots, n.
$$

Differentiating $g(q) = 0$ gives $\nabla g(q)\, dq/dt = \nabla g(q)\, v = 0$; differentiating again gives

$$
\begin{aligned}
0 &= \nabla_q\left(\nabla g(q)\, v \right) \frac{dq}{dt} + \nabla g(q)\frac{dv}{dt} \\
&= \nabla_q\left(\nabla g(q)\, v \right) v + \nabla g(q)\, M(q)^{-1}\left[k(q, v) - \nabla V(q) - \nabla g(q)^T \lambda \right],
\end{aligned}
$$

which can be solved for λ in terms of q and v provided $\nabla g(q)\, M(q)^{-1}\, \nabla g(q)^T$ is nonsingular. So, provided $\nabla g(q)\, M(q)^{-1}\, \nabla g(q)^T$ is nonsingular, the system (10.47)–(10.49) is an index 3 DAE. Since $M(q)$ can usually be taken to be symmetric positive definite, all that is really needed is for $\nabla g(q)$ to have full row rank (i.e., the rows of $\nabla g(q)$ should be linearly independent).

Note that we need initial conditions to be consistent; that is, $g(q(t_0)) = 0$ and

$$
(d/dt)g(q(t))|_{t=t_0} = \nabla g(q(t_0))\, v(t_0) = 0.
$$

Indeed, at every time t, we have $g(q(t)) = 0$ and $\nabla g(q(t))\, v(t) = 0$ for the true solution. We can obtain the consistency condition for λ by differentiating $\nabla g(q(t))\, v(t) = 0$ once again.

10.6.1 Runge–Kutta methods for mechanical index 3 systems

Apart from the index reduction techniques introduced at the start of this chapter, we can apply Runge–Kutta methods directly to the system (10.47)–(10.49). The Runge–Kutta equations are even harder to solve than those for index 2 problems (the condition

Table 10.2 Proven order of accuracy for index 3 problems of type $s \leq 3$ for (10.47)–(10.49)

Method	q	v	λ
Radau IIA	$2s - 1$	s	$s - 1$
Lobatto IIIC	$s + 1$	$s - 1$	$s - 2$

number of the Jacobian matrix in Newton's method grows like $\mathcal{O}(h^{-2})$), but this can be done provided the computed generalized coordinates q_n and generalized velocities v_n are sufficiently close to being consistent ($g(q_n) \approx 0$ and $\nabla g(q_n) v_n \approx 0$), and the newly computed values q_{n+1} and v_{n+1} are also close to being consistent.

The order of accuracy is still not known in general for the Gauss, Radau IIA, and Lobatto IIIC families of Runge–Kutta methods. However, for no more than three stages, this is known for the Radau IIA and Lobatto IIIC methods, and is given in Table 10.2 ([42], [49]).

Again, the order of accuracy of the different components (coordinates, velocities, and constraint forces) are different — and again the winner seems to be the Radau IIA methods (at least up to three stages). Indeed, the three-stage Radau IIA method has been implemented as a FORTRAN 77 code called Radau5, which is available from

http://www.unige.ch/~hairer/software.html

Also available from this website is Radau, another FORTRAN 77 code for Radau IIA methods that can switch between the methods of orders 5, 9, and 13 for DAEs and stiff ODEs.

Numerical results for a fixed stepsize, three-stage Radau IIA method are shown in Figure 10.4 for the index 3 version of the spherical pendulum problem. With $s = 3$ we expect fifth-order convergence for positions, third-order convergence for the velocities, and second-order convergence for the forces. Indeed, the slopes of the graphs in Figure 10.4 are $\approx -4.66, -3.04$, and -2.05 for the positions, velocities, and forces, respectively. This slight drop in the slope from 5 to 4.66 for the position errors is due mainly to the accuracy with which the Runge–Kutta equations are solved, which limits the overall accuracy of the numerical solutions. Otherwise, the theoretical expectations are confirmed by these numerical results.

Other approaches to Runge–Kutta methods for index 3 DAEs from mechanics can be found in [50] for constrained Hamiltonian systems using a pair of Runge–Kutta methods. Essentially one Runge–Kutta method is used for the momentum variables and another for the generalized coordinate variables. The optimal choice of methods for this approach is a combination of Lobatto IIIA and Lobatto IIIB methods.

10.7 HIGHER INDEX DAES

The theory and practice of DAEs become harder as the index increases. Beyond index 3, the complexity of establishing the order of convergence of a method (or *if* a method converges) becomes almost prohibitive for standard approaches such as Runge–Kutta methods. Approaches to these problems can be developed by means of symbolic as well as numerical computation. A survey of approaches to handling high-index DAEs can be found in [26]. Software techniques such as *Automatic Differentiation* [29], [69] can be used instead of symbolic computation (as carried out by *Mathematica*[TM], *Maple*[TM], *Macsyma*[TM], etc.). These approaches take us well outside the scope of this book, but may be useful in handling problems of this kind.

PROBLEMS

1. Obtain the Radau or Radau5 code, and use it to solve the pendulum DAE (10.4)–(10.8) as a DAE.

2. Repeat Problem 1 with the reduced index DAE (10.4)–(10.7) with the constraint $0 = xu + yv$. This is an index 2 DAE. In particular, check the drift, or how far $r^2 + y^2 - l^2$ is from zero.

3. Repeat Problem 1 with the ODE (10.13)–(10.16). As in Problem 3, check the drift in both $x^2 + y^2 - l^2$ and in $xu + yv$ from zero.

4. Repeat Problem 3 using the MATLAB® routine ode23t instead of Radau or Radau5.

5. Consider a system of chemical reactions

$$X + Y \rightarrow Z,$$
$$Y + U \rightleftharpoons V.$$

Assuming that these are *simple* reactions, the reaction rate of the first is proportional to the products of the concentrations of X and Y; that is, for the first reaction, we obtain

$$\frac{d[X]}{dt} = -k_1[X][Y],$$
$$\frac{d[Z]}{dt} = +k_1[X][Y].$$

However, the second reaction is reversible:

$$\frac{d[V]}{dt} = +k_2[Y][U] - k_3[V],$$
$$\frac{d[U]}{dt} = -k_2[Y][U] + k_3[V].$$

Chemical species Y participates in both reactions:

$$\frac{d[Y]}{dt} = +k_3[V] - k_1[X][Y].$$

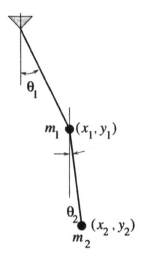

Figure 10.5 Compound pendulum

Suppose that $k_2, k_3 \gg k_1$, enabling us to treat the second reaction as being very nearly in equilibrium. (Mathematically, consider the limit as $k_2, k_3 \to \infty$ but $k_2/k_3 \to c$.) Write down the resulting system of differential and algebraic equations (perhaps involving the initial concentrations $[Y]_0$, $[U]_0$, $[V]_0$, etc.). Show that they form an index 1 DAE.

6. Derive the equations of motion of a compound pendulum as shown in Figure 10.5 as an index 3 DAE in terms of the coordinates of the centers of masses (x_1, y_1) and (x_2, y_2). This will entail the use of two constraints: $x_1^2 + y_1^2 = l_1^2$ and $(x_2 - x_1)^2 + (y_2 - y_1)^2 = l_2^2$. Compare this with the same derivation instead using just two generalized coordinates, θ_1 and θ_2. (Using θ_1 and θ_2 will give ugly expressions for the kinetic energy, but with fewer variables than using x_1, y_1, x_2, and y_2.)

CHAPTER 11

TWO-POINT BOUNDARY VALUE PROBLEMS

In Chapter 3 we saw that the initial value problem for the second-order equation

$$Y'' = f(t, Y, Y') \tag{11.1}$$

can be reformulated as an initial value problem for a system of first-order equations, and that numerical methods for first-order initial value problems can then be applied to this system. In this chapter, we consider the numerical solution of another type of problem for the second-order equation (11.1), one where conditions on the solution Y are given at two distinct t values. Such a problem is called a *two-point boundary value problem* (or sometimes for brevity, a BVP). For simplicity, we begin our discussion with the following BVP for a second-order *linear* equation:

$$Y''(t) = p(t) Y'(t) + q(t) Y(t) + r(t), \quad a < t < b, \tag{11.2}$$

$$Y(a) = g_1, \quad Y(b) = g_2. \tag{11.3}$$

The conditions $Y(a) = g_1$ and $Y(b) = g_2$ are called the *boundary conditions*.

Boundary conditions involving the derivative of the unknown function are also common in applications, and we discuss them later in the chapter.

We assume the given functions p, q and r to be continuous on $[a, b]$. A standard theoretical result states that if $q(t) > 0$ for $t \in [a, b]$, then the boundary value problem

(11.2)–(11.3) has a unique solution; see Keller [53, p. 11]. We will assume that the problem has a unique smooth solution Y.

We begin our discussion of the numerical solution of BVPs by introducing a finite-difference approximation to (11.2). Later we look at more general two-point BVPs for the more general nonlinear second-order equation (11.1), generalizing finite-difference approximations as well. We also introduce other numerical methods for these nonlinear BVPs.

11.1 A FINITE-DIFFERENCE METHOD

The main feature of the finite-difference method is to obtain discrete equations by replacing derivatives with appropriate finite divided differences. We derive a finite-difference system for the BVP (11.2)–(11.3) in three steps.

In the first step, we discretize the domain of the problem: the interval $[a, b]$. Let N be a positive integer, and divide the interval $[a, b]$ into N equal parts:

$$[a, b] = [t_0, t_1] \cup [t_1, t_2] \cup \cdots \cup [t_{N-1}, t_N],$$

where $a = t_0 < t_1 < \cdots < t_{N-1} < t_N = b$ are the grid (or node) points. Denote $h = (b - a)/N$, called the *stepsize*. Then the node points are given by

$$t_i = a + i h, \quad 0 \le i \le N. \tag{11.4}$$

A nonuniform partition of the interval is also possible, and in fact this is preferable if the solution of the boundary value problem (11.2)–(11.3) changes much more rapidly in some parts of $[a, b]$ as compared to other parts of the interval. We restrict our presentation to the case of uniform partitions for the simplicity of exposition. We use the notation $p_i = p(t_i)$, $q_i = q(t_i)$, $r_i = r(t_i)$, $0 \le i \le N$, and denote y_i, $0 \le i \le N$, as numerical approximations of the true solution values $Y_i = Y(t_i)$, $0 \le i \le N$.

In the second step, we discretize the differential equation at the interior node points $t_1, \ldots, t_{N-1}$. For this purpose, let us note the following difference approximation formulas

$$Y'(t_i) = \frac{Y_{i+1} - Y_{i-1}}{2h} - \frac{h^2}{6} Y^{(3)}(\eta_i), \tag{11.5}$$

$$Y''(t_i) = \frac{Y_{i+1} - 2Y_i + Y_{i-1}}{h^2} - \frac{h^2}{12} Y^{(4)}(\xi_i) \tag{11.6}$$

for some $t_{i-1} \le \xi_i, \eta_i \le t_{i+1}$, $i = 1, \ldots, N - 1$. The errors can be obtained by using Taylor polynomial approximations to $Y(t)$. We leave this as an exercise for the reader; or see [11, §5.7], [12, §5.4]. Using these relations, the differential equation at $t = t_i$ becomes

$$\frac{Y_{i+1} - 2Y_i + Y_{i-1}}{h^2} = p_i \frac{Y_{i+1} - Y_{i-1}}{2h} + q_i Y_i + r_i + \mathcal{O}(h^2). \tag{11.7}$$

Dropping the remainder term $\mathcal{O}(h^2)$ and replacing Y_i by y_i, we obtain the difference equations

$$\frac{y_{i+1} - 2\,y_i + y_{i-1}}{h^2} = p_i \frac{y_{i+1} - y_{i-1}}{2\,h} + q_i y_i + r_i, \quad 1 \le i \le N-1, \quad (11.8)$$

which can be rewritten as

$$-\left(1 + \tfrac{1}{2}hp_i\right) y_{i-1} + (2 + h^2 q_i)y_i + \left(\tfrac{1}{2}hp_i - 1\right) y_{i+1}$$
$$= -h^2 r_i, \quad 1 \le i \le N-1. \quad (11.9)$$

The third step is devoted to the treatment of the boundary conditions. The difference equations (11.9) consist of $N-1$ equations for $N+1$ unknowns $y_0, y_1, \ldots, y_N$. We need two more equations, and they come from discretization of the boundary conditions. For the model problem (11.2)–(11.3), the discretization of the boundary conditions is straightforward:

$$y_0 = g_1, \quad y_N = g_2. \quad (11.10)$$

Equations (11.9) and (11.10) together form a linear system. Since the values of y_0 and y_N are explicitly given in (11.10), we can eliminate y_0 and y_N from the linear system. With $y_0 = g_1$, we can rewrite the equation in (11.9) with $i = 1$ as

$$(2 + h^2 q_1)y_1 + \left(\tfrac{1}{2}hp_1 - 1\right) y_2 = -h^2 r_1 + \left(1 + \tfrac{1}{2}hp_1\right) g_1. \quad (11.11)$$

Similarly, from the equation in (11.9) with $i = N-1$, we obtain

$$-\left(1 + \tfrac{1}{2}hp_{N-1}\right) y_{N-2} + (2 + h^2 q_{N-1}) y_{N-1}$$
$$= -h^2 r_{N-1} + \left(1 - \tfrac{1}{2}hp_{N-1}\right) g_2. \quad (11.12)$$

So finally, the finite-difference system for the unknown numerical solution vector $\mathbf{y} = [y_1, \cdots, y_{N-1}]^T$ is

$$A\mathbf{y} = \mathbf{b}, \quad (11.13)$$

where

$$A = \begin{bmatrix} 2 + h^2 q_1 & \tfrac{1}{2}hp_1 - 1 & & & \\ -\left(1 + \tfrac{1}{2}hp_2\right) & 2 + h^2 q_2 & \tfrac{1}{2}hp_2 - 1 & & \\ & \ddots & \ddots & \ddots & \\ & & 2 + h^2 q_{N-2} & \tfrac{1}{2}hp_{N-2} - 1 \\ & & -\left(1 + \tfrac{1}{2}hp_{N-1}\right) & 2 + h^2 q_{N-1} \end{bmatrix}$$

is the coefficient matrix and

$$b_i = \begin{cases} -h^2 r_1 + \left(1 + \tfrac{1}{2}hp_1\right) g_1, & i = 1 \\ -h^2 r_i, & i = 2, \ldots, N-2 \\ -h^2 r_{N-1} + \left(1 - \tfrac{1}{2}hp_{N-1}\right) g_2, & i = N-1. \end{cases} \quad (11.14)$$

The linear system (11.13) is *tridiagonal*, and the solution of tridiagonal linear systems is a very well-studied problem. Examples of programs for the efficient solution of tridiagonal linear systems can be found in *LAPACK* [3].

Table 11.1 Numerical errors $Y(x) - y_h(x)$ for solving (11.19)

t	$h = 1/20$	$h = 1/40$	Ratio	$h = 1/80$	Ratio	$h = 1/160$	Ratio
0.1	5.10e − 5	1.27e − 5	4.00	3.18e − 6	4.00	7.96e − 7	4.00
0.2	7.84e − 5	1.96e − 5	4.00	4.90e − 6 ⋅	4.00	1.22e − 6	4.00
0.3	8.64e − 5	2.16e − 5	4.00	5.40e − 6	4.00	1.35e − 6	4.00
0.4	8.08e − 5	2.02e − 5	4.00	5.05e − 6	4.00	1.26e − 6	4.00
0.5	6.73e − 5	1.68e − 5	4.00	4.21e − 6	4.00	1.05e − 6	4.00
0.6	5.08e − 5	1.27e − 5	4.00	3.17e − 6	4.00	7.94e − 7	4.00
0.7	3.44e − 5	8.60e − 6	4.00	2.15e − 6	4.00	5.38e − 7	4.00
0.8	2.00e − 5	5.01e − 6	4.00	1.25e − 6	4.00	3.13e − 7	4.00
0.9	8.50e − 6	2.13e − 6	4.00	5.32e − 7	4.00	1.33e − 7	4.00

11.1.1 Convergence

It can be shown that if the true solution $Y(t)$ is sufficiently smooth, say, with continuous derivatives up to order 4, then the difference scheme (11.13)–(11.14) is a second-order method,

$$\max_{0 \le i \le N} |Y(t_i) - y_i| = \mathcal{O}(h^2). \tag{11.15}$$

For a detailed discussion, see Ascher et al. [9, p. 189]. Moreover, if $Y(t)$ has six continuous derivatives, the following asymptotic error expansion holds:

$$Y(t_i) - y_h(t_i) = h^2 D(t_i) + \mathcal{O}(h^4), \quad 0 \le i \le N \tag{11.16}$$

for some function $D(t)$ independent of h. The Richardson extrapolation formula for this case is

$$\tilde{y}_h(t_i) = \tfrac{1}{3} \left[4\, y_h(t_i) - y_{2h}(t_i) \right], \tag{11.17}$$

and we have

$$Y(t_i) - \tilde{y}_h(t_i) = \mathcal{O}(h^4). \tag{11.18}$$

11.1.2 A numerical example

We illustrate the finite-difference approximation (11.12), the error result (11.15), and the Richardson extrapolation results (11.16)–(11.18). The MATLAB® codes that we use for our calculations are given following the example.

Example 11.1 Consider the boundary value problem

$$\begin{cases} Y'' = -\dfrac{2t}{1+t^2} Y' + Y + \dfrac{2}{1+t^2} - \log(1+t^2), & 0 < t < 1, \\ Y(0) = 0, \quad Y(1) = \log(2). \end{cases} \tag{11.19}$$

The true solution is $Y(t) = \log(1+t^2)$. In Table 11.1, we report the finite-difference solution errors $Y - y_h$ at selected node points for several values of h. In Table 11.2,

Table 11.2 Extrapolation errors $Y(t_i) - \widetilde{y}_h(t_i)$ for solving (11.19)

t	$h = 1/40$	$h = 1/80$	Ratio	$h = 1/160$	Ratio
0.1	$-9.23e - 09$	$-5.76e - 10$	16.01	$-3.60e - 11$	16.00
0.2	$-1.04e - 08$	$-6.53e - 10$	15.99	$-4.08e - 11$	15.99
0.3	$-6.60e - 09$	$-4.14e - 10$	15.96	$-2.59e - 11$	15.98
0.4	$-1.18e - 09$	$-7.57e - 11$	15.64	$-4.78e - 12$	15.85
0.5	$3.31e - 09$	$2.05e - 10$	16.14	$1.28e - 11$	16.06
0.6	$5.76e - 09$	$3.59e - 10$	16.07	$2.24e - 11$	16.04
0.7	$6.12e - 09$	$3.81e - 10$	16.04	$2.38e - 11$	16.03
0.8	$4.88e - 09$	$3.04e - 10$	16.03	$1.90e - 11$	16.03
0.9	$2.67e - 09$	$1.67e - 10$	16.02	$1.04e - 11$	16.03

we report the errors of the extrapolated solutions $Y - \frac{1}{3}(4\,y_h - y_{2h})$ at the same node points and the associated ratios of the errors for different stepsizes. The column marked "Ratio" next to the column of the solution errors for a stepsize h consists of the ratios of the solution errors for the stepsize $2h$ with those for the stepsize h. We clearly observe an error reduction of a factor of approximately 4 when the stepsize is halved, indicating a second-order convergence of the method as asserted in (11.15).

There is a dramatic improvement in the solution accuracy through extrapolation. The extrapolated solution $\widetilde{y}_h$ with $h = 1/40$ is much more accurate than the solution y_h with $h = 1/160$. Note that the cost of obtaining $\widetilde{y}_h$ with $h = 1/40$ is substantially smaller than that for y_h with $h = 1/160$. Also observe that for the extrapolated solution $\widetilde{y}_h$, the error decreases by a factor of approximately 16 when h is halved. Indeed, it can be shown that if the true solution $Y(t)$ is 8 times continuously differentiable, then we can improve the asymptotic error expansion (11.16) to

$$Y(t_i) - y_h(t_i) = h^2 D_1(t_i) + h^4 D_2(t_i) + \mathcal{O}(h^6). \tag{11.20}$$

Then (11.17) is replaced by

$$Y(t_i) - \widetilde{y}_h(t_i) = -4\,h^4 D_2(t_i) + \mathcal{O}(h^6). \tag{11.21}$$

Therefore, we can also perform an extrapolation procedure on $\widetilde{y}_h$ to get an even more accurate numerical solution through the following formula:

$$Y(t_i) - \frac{1}{15}\left[16\,\widetilde{y}_h(t_i) - \widetilde{y}_{2h}(t_i)\right] = \mathcal{O}(h^6). \tag{11.22}$$

As an example, at $t_i = 0.5$, with $h = 1/80$, the doubly extrapolated solution has an error approximately equal to -1.88×10^{-12}.   ∎

MATLAB program. The following MATLAB code ODEBVP implements the difference method (11.13) for solving the problem (11.2)–(11.3).

```
function z = ODEBVP(p,q,r,a,b,ga,gb,N)
%
```

```
% function z = ODEBVP(p,q,r,a,b,ga,gb,N)
%
% A program to solve the two point boundary
% value problem
%    y"=p(t)y'+q(t)y+r(t),   a<t<b
%    y(a)=g1,   y(b)=g2
% Input
%    p, q, r: coefficient functions
%    a, b: the end-points of the interval
%    ga, gb: the prescribed function values
%            at the end-points
%    N: number of sub-intervals
% Output
%    z = [ tt yy ]: tt is an (N+1) column vector
%                   of the node points
%                   yy is an (N+1) column vector of
%                   the solution values
% A sample call would be
%    z=ODEBVP('p','q','r',a,b,ga,gb,100)
% The user must provide m-files to define the
% functions p, q, and r.
%
% The user must also supply a MATLAB program, called
% tridiag.m, for solving tridiagonal linear systems.
%
% Initialization
N1 = N+1;
h = (b-a)/N;
h2 = h*h;
tt = linspace(a,b,N1)';
yy = zeros(N1,1);
yy(1) = ga;
yy(N1) = gb;
% Define the sub-diagonal avec, main diagonal bvec,
% superdiagonal cvec
pp(2:N) = feval(p,tt(2:N));
avec(2:N-1) = -1-(h/2)*pp(3:N);
bvec(1:N-1) = 2+h2*feval(q,tt(2:N));
cvec(1:N-2) = -1+(h/2)*pp(2:N-1);
% Define the right hand side vector fvec
fvec(1:N-1) = -h2*feval(r,tt(2:N));
fvec(1) = fvec(1)+(1+h*pp(2)/2)*ga;
fvec(N-1) = fvec(N-1)+(1-h*pp(N)/2)*gb;
% Solve the tridiagonal system
yy(2:N) = tridiag(avec,bvec,cvec,fvec,N-1,0);
```

```
z = [tt'; yy']';
```

The following MATLAB code tridiag solves tridiagonal linear systems.

```
function [x, alpha, beta, message] = tridiag(a,b,c,f,n,option)
%
% function [x, alpha, beta, message] = tridiag(a,b,c,f,n,option)
%
% Solve a tridiagonal linear system M*x=f
%
% INPUT:
% The order of the linear system is given as n.
% The subdiagonal, diagonal, and superdiagonal of M are given
% by the arrays a,b,c, respectively.  More precisely,
%     M(i,i-1) = a(i), i=2,...,n
%     M(i,i)  = b(i), i=1,...,n
%     M(i,i+1) = c(i), i=1,...,n-1
% option=0 means that the original matrix M is given as
%     specified above.
% option=1 means that the LU factorization of M is already
%     known and is stored in a,b,c.  This will have been
%     accomplished by a previous call to this routine.  In
%     that case, the vectors alpha and beta should have
%     been substituted for a and b in the calling sequence.
% All input values are unchanged on exit from the routine.
%
% OUTPUT:
% Upon exit, the LU factorization of M is already known and
% is stored in alpha,beta,c.  The solution x is given as well.
% message=0 means the program was completed satisfactorily.
% message=1 means that a zero pivot element was encountered
%     and the solution process was abandoned.  This case
%     happens only when option=0.

if option == 0
 alpha = a; beta = b;
 alpha(1) = 0;

 % Compute LU factorization of matrix M.
 for j=2:n
     if beta(j-1) == 0
         message = 1; return
     end
     alpha(j) = alpha(j)/beta(j-1);
     beta(j) = beta(j) - alpha(j)*c(j-1);
 end
```

```
    if beta(n) == 0
        message = 1; return
    end
end

% Compute solution x to M*x = f using LU factorization of M.
% Do forward substitution to solve lower triangular system.
if option == 1
        alpha = a; beta = b;
end
x = f; message = 0;

for j=2:n
        x(j) = x(j) - alpha(j)*x(j-1);
end

% Do backward substitution to solve upper triangular system.
x(n) = x(n)/beta(n);
for j=n-1:-1:1
        x(j) = (x(j) - c(j)*x(j+1))/beta(j);
end

end % tridiag
```

11.1.3 Boundary conditions involving the derivative

The treatment of boundary conditions involving the derivative of the unknown $Y(t)$ is somewhat involved. Assume that the boundary condition at $t = b$ is

$$Y'(b) + k\,Y(b) = g_2. \tag{11.23}$$

One obvious discretization is to approximate $Y'(b)$ by $(Y_N - Y_{N-1})/h$. However,

$$Y'(b) - \frac{Y_N - Y_{N-1}}{h} = \mathcal{O}(h), \tag{11.24}$$

and the accuracy of this approximation is one order lower than the remainder term $\mathcal{O}(h^2)$ in (11.7). As a result, the corresponding difference solution with the following discrete boundary condition

$$\frac{y_N - y_{N-1}}{h} + k\,y_N = g_2 \tag{11.25}$$

will have an accuracy of $\mathcal{O}(h)$ only. To retain the second-order convergence of the difference solution, we need to approximate the boundary condition (11.23) more accurately. One such treatment is based on the formula

$$Y'(b) = \frac{3\,Y_N - 4\,Y_{N-1} + Y_{N-2}}{2\,h} + \mathcal{O}(h^2). \tag{11.26}$$

Then the boundary condition (11.23) is approximated by

$$\frac{3\,y_N - 4\,y_{N-1} + y_{N-2}}{2\,h} + k\,y_N = g_2. \tag{11.27}$$

It can be shown that the resulting difference scheme is again second-order accurate.

A similar treatment can be given for more general boundary conditions that involve the derivatives $Y'(a)$ and $Y'(b)$. For a comprehensive introduction to this and to the general subject of the numerical solution of two-point boundary value problems, see Keller [53], Ascher et al [9], or Ascher and Petzold [10, Chap. 6].

11.2 NONLINEAR TWO-POINT BOUNDARY VALUE PROBLEMS

Consider the two-point boundary value problem

$$Y'' = f(t, Y, Y'), \qquad a < t < b,$$

$$A \begin{bmatrix} Y(a) \\ Y'(a) \end{bmatrix} + B \begin{bmatrix} Y(b) \\ Y'(b) \end{bmatrix} = \begin{bmatrix} \gamma_1 \\ \gamma_2 \end{bmatrix}. \tag{11.28}$$

The terms A and B denote given square matrices of order 2×2, and γ_1 and γ_2 are given constants. The theory for BVPs such as this one is more complex than that for the initial value problem.

The theory for the nonlinear problem (11.28) is more complicated than that for the linear problem (11.2). We give an introduction to that theory for the following more limited problem:

$$Y'' = f(t, Y, Y'), \qquad a < t < b, \tag{11.29}$$

$$a_0 y(a) - a_1 y'(a) = g_1, \qquad b_0 y(b) + b_1 y'(b) = g_2 \tag{11.30}$$

with $\{a_0, a_1, b_0, b_1, g_1, g_2\}$ as given constants. The function f is assumed to satisfy the following Lipschitz condition,

$$|f(t, u_1, v) - f(t, u_2, v)| \le K\,|u_1 - u_2|,$$

$$|f(t, u, v_1) - f(t, u, v_2)| \le K\,|v_1 - v_2| \tag{11.31}$$

for all points (t, u_i, v), (t, u, v_i), $i = 1, 2$, in the region

$$R = \{(t, u, v) \mid a \le t \le b, \ -\infty < u, v < \infty\}.$$

This is far stronger than needed, but it simplifies the statement of the following theorem; and although we do not give it here, it also simplifies the error analysis of numerical methods for (11.29)–(11.30).

Theorem 11.2 *For the problem (11.29)–(11.30), assume $f(x, u, v)$ to be continuous on the region R and that it satisfies the Lipschitz condition (11.31). In addition, assume that on R, f satisfies*

$$\frac{\partial f(x, u, v)}{\partial u} > 0, \qquad \left| \frac{\partial f(x, u, v)}{\partial v} \right| \le M \tag{11.32}$$

for some constant $M > 0$. For the boundary conditions of (11.30), assume

$$a_0 a_1 \geq 0, \qquad b_0 b_1 \geq 0, \tag{11.33}$$

$$|a_0| + |a_1| \neq 0, \qquad |b_0| + |b_1| \neq 0, \qquad |a_0| + |b_0| \neq 0.$$

Then the BVP (11.29)–(11.30) has a unique solution.

For a proof, see Keller [53, p. 9].

Although this theorem gives conditions for the BVP (11.29)–(11.30) to be uniquely solvable, in fact nonlinear BVPs may be nonuniquely solvable with only a finite number of solutions. This is in contrast to the situation for linear problems such as (11.2)–(11.3) in which nonuniqueness always implies an infinity of solutions. An example of such nonunique solvability for a nonlinear BVP is the second-order problem

$$\frac{d}{dt}\left[I(t)\frac{dY}{dt}\right] + \lambda \sin(Y) = 0, \qquad 0 < t < 1,$$

$$Y'(0) = Y'(1) = 0, \qquad |Y(t)| < \pi, \tag{11.34}$$

which arises in studying the buckling of a vertical column when a vertical force is applied. The unknown $Y(t)$ is related to the displacement of the column in the radial direction from its centerline. In the equation $I(t)$ is a given function related to physical properties of the column; and the parameter λ is proportional to the load on the column. When λ exceeds a certain size, there is a solution to the problem (11.34) other than the zero solution. As λ continues to increase, the BVP (11.34) has an increasing number of nonzero solutions, only one of which is the correct physical solution. For a detailed discussion of this problem, see Keller and Antman [54, p. 43].

As with the earlier material on initial value problems in Chapter 3, all boundary value problems for higher-order equations can be reformulated as problems for systems of first-order equations. The general form of a two-point BVP for a system of first-order equations is

$$\mathbf{Y}' = \mathbf{f}(t, \mathbf{Y}), \qquad a < t < b,$$

$$A\mathbf{Y}(a) + B\mathbf{Y}(b) = \mathbf{g}. \tag{11.35}$$

This represents a system of m first-order equations. The quantities $\mathbf{Y}(t)$, $\mathbf{f}(t, \mathbf{Y})$, and g are vectors with m components, and A and B are matrices of order $m \times m$. There is a theory for such BVPs, analogous to that for the two-point problem (11.28), but we omit it here because of space limitations.

In the remainder of this section, we describe briefly the principal numerical methods for solving the two-point BVP (11.28). These methods generalize to first-order systems such as (11.35), but again, because of space limitations, we omit those results. Much of our presentation follows Keller [53], and a theory for first-order systems is given there. Unlike the situation with initial value problems, it is often advantageous to directly treat higher-order BVPs rather than to numerically solve their reformulation as a first-order system. The numerical methods for the two-point boundary value problem (11.28) are also less complicated to present, and therefore we have opted to discuss the second-order problem (11.28) rather than the system (11.35).

11.2.1 Finite difference methods

We consider the two-point BVP:

$$Y'' = f(t, Y, Y'), \quad a < t < b,$$
$$Y(a) = g_1, \quad Y(b) = g_2.$$
(11.36)

with the true solution denoted by $Y(t)$. The boundary conditions are of the same form as used with our earlier finite-difference approximation for the linear problem (11.2)–(11.3). As before, in (11.4), introduce an equally spaced subdivision

$$a = t_0 < t_1 < \cdots < t_N = b$$

At each interior node point t_i, $0 < i < N$, we approximate $Y''(t_i)$ and $Y'(t_i)$ as in (11.5)–(11.6). Dropping the final error terms in (11.5)–(11.6) and using these approximations in the differential equation, we are led to the approximating nonlinear system:

$$\frac{y_{i+1} - 2y_i + y_{i-1}}{h^2} = f\left(t_i, y_i, \frac{y_{i+1} - y_{i-1}}{2h}\right), \quad i = 1, \ldots, N - 1. \quad (11.37)$$

This is a system of $N - 1$ nonlinear equations in the $N - 1$ unknowns $y_1, \ldots, y_{N-1}$; compare with the system (11.8). The values $y_0 = g_1$ and $y_N = g_2$ are known from the boundary conditions.

The analysis of the error in $\{y_i\}$ as compared to $\{Y(t_i)\}$ is too complicated to be given here, because it requires methods for analyzing the solvability of systems of nonlinear equations. In essence, if $Y(t)$ is 4 times differentiable, if the problem (11.36) is uniquely solvable for some region about the graph on $[a, b]$ of $Y(t)$, and if $f(t, u, v)$ is sufficiently differentiable, then there is a solution to (11.37), and it satisfies

$$\max_{0 \le i \le N} |Y(t_i) - y_i| = \mathcal{O}(h^2). \quad (11.38)$$

For an analysis, see Keller [52, Sec. 3.2] or [53, Sec. 3.2]. Moreover, with additional assumptions on f and the smoothness of Y, it can be shown that

$$Y(t_i) - y_i = D(t_i)h^2 + \mathcal{O}(h^4) \quad (11.39)$$

with $D(t)$ independent of h. This can be used to justify Richardson extrapolation to obtain results that converge more rapidly, just as earlier in (11.16)–(11.18). (There are other methods for improving the convergence, based on correcting for the error in the central difference approximations of (11.5)–(11.6); e.g., see [27], [77].)

The system (11.37) can be solved in a variety of ways, some of which are simple modifications of Newton's method for solving systems of nonlinear equations. We describe here the application of the standard Newton method.

In matrix form, we have

$$
\frac{1}{h^2}
\begin{bmatrix}
-2 & 1 & 0 & & \cdots & 0 \\
1 & -2 & 1 & & & \vdots \\
\vdots & & \ddots & & & \\
& & & 1 & -2 & 1 \\
0 & \cdots & & 0 & 1 & -2
\end{bmatrix}
\begin{bmatrix}
y_1 \\
y_2 \\
\vdots \\
\\
y_{N-1}
\end{bmatrix}
$$

$$
=
\begin{bmatrix}
f\left(t_1, y_1, \frac{1}{2h}(y_2 - g_1)\right) \\
f\left(t_2, y_2, \frac{1}{2h}(y_3 - y_1)\right) \\
\vdots \\
f\left(t_{N-1}, y_{N-1}, \frac{1}{2h}(g_2 - y_{N-2})\right)
\end{bmatrix}
-
\begin{bmatrix}
\dfrac{g_1}{h^2} \\
0 \\
\vdots \\
\dfrac{g_2}{h^2}
\end{bmatrix},
$$

which we denote by

$$
\frac{1}{h^2}T\mathbf{y} = \hat{\mathbf{f}}(\mathbf{y}) + \mathbf{g}. \tag{11.40}
$$

The matrix T is both tridiagonal and nonsingular (see Problem 14). As was discussed earlier for the solution of (11.13) for the linear BVP (11.2)–(11.3), tridiagonal linear systems $T\mathbf{z} = \mathbf{b}$ are easily solvable. This can be used to show that (11.40) is solvable for all sufficiently small values of h; moreover, the solution is unique in a region of $\mathbb{R}^{N-1}$ corresponding to some neighborhood of the graph of the solution $Y(t)$ for the original BVP (11.36). Newton's method (see [11, §2.11]) for solving (11.40) is given by

$$
\mathbf{y}^{(m+1)} = \mathbf{y}^{(m)} - \left[\frac{1}{h^2}T - F(\mathbf{y}^{(m)})\right]^{-1}\left[\frac{1}{h^2}T\mathbf{y}^{(m)} - \hat{\mathbf{f}}(\mathbf{y}^{(m)}) - \mathbf{g}\right] \tag{11.41}
$$

with F the Jacobian matrix for $\hat{\mathbf{f}}$,

$$
F(\mathbf{y}) = \left[\frac{\partial \hat{f}_i}{\partial y_j}\right]_{i,j=1,\ldots,N-1}
$$

This matrix simplifies considerably because of the special form of $\hat{\mathbf{f}}(\mathbf{y})$,

$$
[F(\mathbf{y})]_{ij} = \frac{\partial}{\partial y_j}f\left(t_i, y_i, \frac{1}{2h}(y_{i+1} - y_{i-1})\right).
$$

This is zero unless $j = i - 1, i$, or $i + 1$:

$$[F(\mathbf{y})]_{ii} = f_2\left(t_i, y_i, \frac{1}{2h}(y_{i+1} - y_{i-1})\right), \qquad 1 \leq i \leq N - 1,$$

$$[F(\mathbf{y})]_{i,i-1} = \frac{-1}{2h} f_3\left(t_i, y_i, \frac{1}{2h}(y_{i+1} - y_{i-1})\right), \qquad 2 \leq i \leq N - 1,$$

$$[F(\mathbf{y})]_{i,i+1} = \frac{1}{2h} f_3\left(t_i, y_i, \frac{1}{2h}(y_{i+1} - y_{i-1})\right), \qquad 1 \leq i \leq N - 2$$

with $f_2(t, u, v)$ and $f_3(t, u, v)$ denoting partial derivatives of f with respect to u and v, respectively. Thus the matrix being inverted in (11.41) is tridiagonal. Letting

$$B_m = \frac{1}{h^2} T - F(\mathbf{y}^{(m)}), \tag{11.42}$$

we can rewrite (11.41) as

$$\mathbf{y}^{(m+1)} = \mathbf{y}^{(m)} - \delta^{(m)},$$
$$B_m \delta^{(m)} = \frac{1}{h^2} T \mathbf{y}^{(m)} - \mathbf{f}(\mathbf{y}^{(m)}) - \mathbf{g}. \tag{11.43}$$

This linear system is easily and rapidly solvable, for example, using the MATLAB code of Subsection 11.1.2. The number of multiplications and divisions can be shown to equal approximately $5N$, a relatively small number of operations for solving a linear system of $N - 1$ equations. Additional savings can be made by not varying B_m or by changing it only after several iterations of (11.43). For an extensive survey and discussion of the solution of nonlinear systems that arise in connection with solving BVPs, see Deuflhard [32].

Example 11.3 Consider the two-point BVP:

$$Y'' = -y + \frac{2(Y')^2}{Y}, \qquad -1 < x < 1, \tag{11.44}$$
$$Y(-1) = Y(1) = (e + e^{-1})^{-1} \doteq 0.324027137.$$

The true solution is $Y(t) = (e^t + e^{-t})^{-1}$. We applied the preceding finite-difference procedure (11.37) to the solution of this BVP. The results are given in Table 11.3 for successive doublings of $N = 2/h$. The nonlinear system in (11.37) was solved using Newton's method, as described in (11.43). The initial guess was

$$y_h^{(0)}(x_i) = (e + e^{-1})^{-1}, \qquad i = 0, 1, \ldots, N,$$

based on connecting the boundary values by a straight line. The quantity

$$d_h = \max_{0 \leq i \leq N} \left| y_i^{(m+1)} - y_i^{(m)} \right|$$

Table 11.3   Finite difference method for solving (11.44)

$N = 2/h$	E_h	Ratio
4	$2.63e-2$	
8	$5.87e-3$	4.48
16	$1.43e-3$	4.11
32	$3.55e-4$	4.03
64	$8.86e-5$	4.01

was computed for each iterate, and when the condition

$$d_h \leq 10^{-10}$$

was satisfied, the iteration was terminated. In all cases, the number of iterates computed was 5 or 6. For the error, let

$$E_h = \max_{0 \leq i \leq N} |Y(x_i) - y_h(x_i)|$$

with y_h the solution of (11.37) obtained with Newton's method. According to (11.38) and (11.39), we should expect the values E_h to decrease by a factor of approximately 4 when h is halved, and that is what we observe in the table. ∎

Higher-order methods can be obtained in several ways.

1. Using higher-order approximations to the derivatives, improving (11.5)–(11.6).

2. Using Richardson extrapolation based on (11.39), as was done in Subsection 11.1.1 for the linear BVP (11.2)–(11.3). Richardson extrapolation can be used repeatedly to obtain methods of increasingly higher-order. This was discussed in Subsection 11.1.2, yielding the formulas (11.20)–(11.22) for extrapolating twice.

3. The truncation errors in (11.5)–(11.6) can be approximated with higher-order differences using the calculated values of y_h. Using these values as corrections in (11.37), we can obtain a new, more accurate approximation to the differential equation in (11.36), leading to a more accurate solution. This is sometimes called the *method of deferred corrections*; for more recent work, see [27], [77].

All of these techniques have been used, and some have been implemented as quite sophisticated computer codes.

11.2.2 Shooting methods

Another popular approach to solving a two-point BVP is to reduce it to a problem
in which a program for solving initial value problems can be used. We now develop
such a method for the BVP (11.29)–(11.30).

Consider the initial value problem

$$Y'' = f(t, Y, Y'), \qquad a < t < b,$$
$$Y(a) = a_1 s - c_1 g_1 \qquad Y'(a) = a_0 s - c_0 g_1,$$

(11.45)

depending on the parameter s, where c_0 and c_1 are arbitrary (user chosen) constants
satisfying

$$a_1 c_0 - a_0 c_1 = 1.$$

Denote the solution of (11.45) by $Y(t; s)$. Then it is a straightforward calculation
using the initial condition in (11.45) to show that

$$a_0 Y(a; s) - a_1 Y'(a; s) = y_1$$

for all s for which Y exists. This shows that $Y(t; s)$ satisfies the first boundary
condition in (11.30).

Since Y is a solution of (11.29), all that is needed for it to be a solution of the
BVP (11.29)–(11.30) is to have it satisfy the remaining boundary condition at b. This
means that $Y(t; s)$ must satisfy

$$\varphi(s) \equiv b_0 Y(b; s) + b_1 Y'(b; s) - g_2 = 0.$$

(11.46)

This is a nonlinear equation for s. If s^* is a root of $\varphi(s)$, then $Y(t; s^*)$ will satisfy the
BVP (11.29)–(11.30). It can be shown that under suitable assumptions on f and its
boundary conditions, equation (11.46) will have a unique solution s^*; see Keller [53,
p. 9]. We can use a rootfinding method for nonlinear equations to solve for s^*. This
way of finding a solution to a BVP is called a *shooting method*. The name comes
from ballistics, in which one attempts to determine the needed initial conditions at
$t = a$ in order to obtain a certain value at $t = b$.

Most rootfinding methods can be applied to solving $\varphi(s) = 0$. Each evaluation
of $\varphi(s)$ involves the solution of the initial value problem (11.45) over $[a, b]$, and
consequently, we want to minimize the number of such evaluations. As a specific
example of an important and rapidly convergent method, we look at Newton's method:

$$s_{m+1} = s_m - \frac{\varphi(s_m)}{\varphi'(s_m)}, \qquad m = 0, 1, \ldots. $$

(11.47)

To calculate $\varphi'(s)$, differentiate the definition (11.46) to obtain

$$\varphi'(s) = b_0 \xi_s(b) + b_1 \xi_s'(b),$$

(11.48)

where

$$\xi_s(t) = \frac{\partial Y(t; s)}{\partial s}.$$

(11.49)

To find $\xi_s(t)$, differentiate the equation

$$Y''(t; s) = f(t, Y(t; s), Y'(t; s))$$

with respect to s. Then ξ_s satisfies the initial value problem

$$\xi_s''(t) = f_2(t, Y(t; s), Y'(t; s))\xi_s(t) + f_3(t, Y(t; s), Y'(t; s))\xi_s'(t), \qquad (11.50)$$

$$\xi_s(a) = a_1, \qquad \xi_s'(a) = a_0.$$

The functions f_2 and f_3 denote the partial derivatives of $f(t, u, v)$ with respect to u and v, respectively. The initial values are obtained from those in (11.45) and from the definition of ξ_s.

In practice, we convert the problems (11.45) and (11.50) to a system of four first-order equations with the unknowns Y, Y', ξ_s, and ξ_s'. This system is solved numerically, say, with a method of order p and stepsize h. Let $y_h(t; s)$ denote the approximation to $Y(t; s)$ with a similar notation for the remaining unknowns. From earlier results for solving initial value problems, it can be shown that these approximate solutions will be in error by $\mathcal{O}(h^p)$. With suitable assumptions on the original problem (11.29)–(11.30), it can then be shown that the root s_h^* obtained will also be in error by $\mathcal{O}(h^p)$ and similarly for the approximate solution $y_h(t; s_h^*)$ when compared to the solution $Y(t; s^*)$ of the boundary value problem. For details of this analysis, see Keller [53, pp. 47–54].

Example 11.4 We apply the preceding shooting method to the solution of the BVP (11.45), used earlier to illustrate the finite-difference method. The initial value problem (11.35) for the shooting method is

$$Y'' = -Y + \frac{2(Y')^2}{Y}, \qquad -1 < x \le 1,$$
$$Y(-1) = (e + e^{-1})^{-1}, \qquad Y'(-1) = s. \qquad (11.51)$$

The associated problem (11.50) for $\xi_s(x)$ is

$$\xi_s'' = \left[-1 - 2\left(\frac{Y'}{Y}\right)^2\right]\xi_s + 4\frac{Y'}{Y}\xi_s', \qquad (11.52)$$
$$\xi_s(-1) = 0, \qquad \xi_s'(-1) = 1.$$

The equation for ξ_s'' uses the solution $Y(x; s)$ of (11.51). The function $\varphi(s)$ for computing s^* is given by

$$\varphi(s) \equiv Y(1; s) - (e + e^{-1})^{-1}.$$

For use in defining Newton's method, we have

$$\varphi'(s) = \xi_s(1).$$

Table 11.4 Shooting method for solving (11.44)

$n = 2/h$	$s^* - s_h^*$	Ratio	E_h	Ratio
4	$4.01e - 3$		$2.83e - 2$	
8	$1.52e - 3$	2.64	$7.30e - 3$	3.88
16	$4.64e - 4$	3.28	$1.82e - 3$	4.01
32	$1.27e - 4$	3.64	$4.54e - 4$	4.01
64	$3.34e - 5$	3.82	$1.14e - 4$	4.00

From the true solution Y of (11.44) and the condition $y'(-1) = s$ in (11.51), the desired root s^* of $\varphi(s)$ is simply

$$s^* = Y'(-1) = \frac{e - e^{-1}}{(e + e^{-1})^2} \doteq 0.245777174.$$

To solve the initial value problem (11.51)–(11.52), we use a second-order Runge–Kutta method, such as (5.21), with a stepsize of $h = 2/n$. The results for several values of n are given in Table 11.4. The solution of (11.52) is denoted by $y_h(t; s)$, and the resulting root for

$$\varphi_h(s) \equiv y_h(1; s) - (e + e^{-1})^{-1} = 0$$

is denoted by s_h^*. For the error in $y_h(t; s_h^*)$, let

$$E_h = \max_{0 \le i \le n} |Y(t_i) - y_h(t_i; s_h^*)|,$$

where $\{t_i\}$ are the node points used in solving the initial value problem. The columns labeled "Ratio" give the factors by which the errors decreased when n was doubled (or h was halved). Theoretically these factors should approach 4 since the Runge–Kutta method has an error of $\mathcal{O}(h^2)$. Empirically, the factors approach 4.0, as expected. For the Newton iteration (11.47), $s_0 = 0.2$ was used in each case. The iteration was terminated when the test

$$|s_{m+1} - s_m| \le 10^{-10}$$

was satisfied. With these choices, the Newton method needed six iterations in each case, except that of $n = 4$ (when seven iterations were needed). However, if $s_0 = 0$ was used, then 25 iterations were needed for the $n = 4$ case, showing the importance of a good choice of the initial guess s_0. ∎

A number of problems can arise with the shooting method. First, there is no general guess s_0 for the Newton iteration, and with a poor choice, the iteration may diverge. For this reason, a modified Newton method may be needed to force convergence. A second problem is that the choice of $y_h(t; s)$ may be very sensitive to h, s, and other characteristics of the boundary value problem. For example, if the linearization of the initial value problem (11.45) has large positive eigenvalues, then the choice of $Y(t; s)$ is likely to be sensitive to variations in s. For a thorough discussion of these and other problems, see Keller [53, Chap. 2], Ascher et al. [9], or Ascher and Petzold [10, Chap. 7]. Some of these problems are more easily examined for linear BVPs, as is done in Keller [53, Chap. 2].

11.2.3 Collocation methods

To simplify the presentation, we again consider only the differential equation

$$Y'' = f(t, Y, Y'), \qquad a < t < b. \tag{11.53}$$

Further simplifying the BVP, we consider only the homogeneous boundary conditions

$$Y(a) = 0, \qquad Y(b) = 0. \tag{11.54}$$

It is straightforward to modify the nonhomogeneous boundary conditions of (11.36) to obtain a modified BVP having homogeneous boundary conditions; see Problem 16. The collocation methods are much more general than indicated by solving (11.53)–(11.54), but the essential ideas are more easily understood in this context.

We assume that the solution $Y(t)$ of (11.53)–(11.54) is approximable by a linear combination of n given functions $\psi_1(t), \ldots, \psi_n(t)$,

$$Y(x) \approx y_n(x) = \sum_{j=1}^{n} c_j \psi_j(t), \qquad a \le x \le b. \tag{11.55}$$

The functions $\psi_j(t)$ are all assumed to satisfy the boundary conditions

$$\psi_j(a) = \psi_j(b) = 0, \qquad j = 1, \ldots, n, \tag{11.56}$$

and thus any linear combination (11.55) will also satisfy the boundary conditions. The coefficients $c_1, \ldots, c_n$ are determined by requiring the differential equation (11.53) to be satisfied exactly at n preselected points in (a, b),

$$y_n''(\xi_i) = f(\xi_i, y_n(\xi_i), y_n'(\xi_i)), \qquad i = 1, \ldots, n \tag{11.57}$$

with given points

$$a < \xi_1 < \xi_2 < \cdots < \xi_n < b. \tag{11.58}$$

The procedure of defining $y_n(t)$ implicitly through (11.57) is known as *collocation*, and the points $\{\xi_i\}$ are called *collocation points*.

Substituting from (11.55) into (11.57), we obtain

$$\sum_{j=1}^{n} c_j \psi_j''(\xi_i) = f\left(\xi_i, \sum_{j=1}^{n} c_j \psi_j(\xi_i), \sum_{j=1}^{n} c_j \psi_j'(\xi_i)\right), \tag{11.59}$$

for $i = 1, \ldots, n$. This is a system of n nonlinear equations in the n unknowns $c_1, \ldots, c_n$. In general, this system must be solved numerically, as is done with the finite-difference approximation (11.37) discussed earlier in Section 11.2.1.

In choosing a collocation method, we must do the following.

1. Choose the family of approximating functions $\{\psi_1(t), \ldots, \psi_n(t)\}$, including the requirement (11.56) for the endpoint boundary conditions.

2. Choose the collocation node points $\{\xi_i\}$ of (11.58).

3. Choose a way to solve the nonlinear system (11.59). Included in this is choosing an initial guess for the method of solving the nonlinear system, and this may be difficult to find.

For a general survey of this area, see the text by Ascher et al. [9]; for collocation software, see [6], [7].

We describe briefly a particular collocation method that has been implemented as a high quality computer code. Let $m > 0$, $h = (b - a)/m$, and define breakpoints $\{t_j\}$ by

$$t_j = a + jh, \quad j = 0, 1, \ldots, m.$$

Consider all functions $p(t)$ that satisfy the following conditions:

- $p(t)$ is continuously differentiable for $a \le t \le b$.

- $p(a) = p(b) = 0$.

- On each subinterval $[t_{j-1}, t_j]$, $p(t)$ is a polynomial of degree ≤ 3.

We use these functions as our approximations $y_n(t)$ in (11.57). There are a number of ways to write $y_n(t)$ in the form of (11.55), with $n = km$. A good way to choose the functions $\{\psi_j(t)\}$ is to use the standard basis functions for cubic Hermite interpolation on each subinterval $[t_{j-1}, t_j]$; see [11, p. 162].

For the collocation points, let $\rho_1 = -1/\sqrt{3}$, $\rho_2 = 1/\sqrt{3}$, which are the zeros of the Legendre polynomial of degree 2 on $[-1, 1]$. Using these, define

$$\xi_{i,j} = \tfrac{1}{2}(t_{i-1} + t_i) + \tfrac{1}{2}h\rho_j, \quad j = 1, 2, \quad i = 1, \ldots, m.$$

This defines $n = 2m$ points $\xi_{i,j}$, and these will be the collocation points used in (11.57).

With this choice for $y_n(t)$ and $\{\xi_{i,j}\}$, and assuming sufficient differentiability and stability in the solvability of the BVP (11.53)–(11.54), it can be shown that $y_n(t)$ satisfies the following:

$$\max_{a \le t \le b} |Y(t) - y_n(t)| = \mathcal{O}(h^4).$$

An extensive discussion and generalizations of this method are given in [9].

11.2.4 Other methods and problems

Yet another approach to solving a boundary value problem is to solve an equivalent reformulation as an integral equation. There is much less development of such numerical methods, although they can be very effective in some situations. For an introduction to this approach, see Keller [53, Chap. 4].

There are also many other types of boundary value problems, some containing certain types of singular behavior, that we have not discussed here. An excellent general reference is the book by Ascher, Mattheij, and Russell [9]. In addition, see the research papers in the proceedings of Ascher and Russell [8], Aziz [13], Childs et al. [28], and Gladwell and Sayers [41]; see also Keller [52, Chap. 4] for singular problems. For discussions of software, see Childs et al. [28], Gladwell and Sayers [41], and Enright [35].

PROBLEMS

1. In general, study of existence and uniqueness of a solution for boundary value problems is more complicated. Consider the boundary value problem

$$\begin{cases} Y''(t) = 0, & 0 < t < 1, \\ Y'(0) = g_1, \ Y'(1) = g_2. \end{cases}$$

   Show that the problem has no solution if $g_1 \neq g_2$, and infinitely many solutions when $g_1 = g_2$.

   Hint: For the case $g_1 \neq g_2$, integrate the differential equation over $[0, 1]$.

2. As another example of solution non-uniqueness, verify that for any constant c, $Y(t) = c \sin(t)$ solves the boundary value problem

$$\begin{cases} Y''(t) + Y(t) = 0, & 0 < t < \pi, \\ Y(0) = Y(\pi) = 0. \end{cases}$$

3. Verify that any function of the form $Y(t) = c_1 e^t + c_2 e^{-t}$ satisfies the equation

$$Y''(t) - Y(t) = 0.$$

   Determine c_1 and c_2 for the function $Y(t)$ to satisfy the following boundary conditions:

   (a) $Y(0) = 1, Y(1) = 0$.
   (b) $Y(0) = 1, Y'(1) = 0$.
   (c) $Y'(0) = 1, Y(1) = 0$.
   (d) $Y'(0) = 1, Y'(1) = 0$.

4. Assume that Y is 3 times continuously differentiable. Use Taylor's theorem to prove the formula (11.26).

5. Prove the formula (11.18) by using the asymptotic expansion (11.16).

6. Use the asymptotic error formula (11.16) with $D(t)$ twice continuously differentiable to show

$$Y''(t_i) - \frac{1}{h^2}\left[y_h(t_{i+1}) - 2y_h(t_i) + y_h(t_{i-1})\right] = \mathcal{O}(h^2), \qquad 1 \le i \le N - 1.$$

In other words, the second-order centered divided difference of the numerical solution is a second-order approximation of the second derivative of the true solution at any interior node point.

7. Verify that any function of the form $Y(t) = c_1\sqrt{t} + c_2 t^4$ satisfies the equation

$$t^2 Y''(t) - \tfrac{7}{2}t Y'(t) + 2Y(t) = 0.$$

Determine the solution of the equation with the boundary conditions

$$Y(1) = 1, \quad Y(4) = 2.$$

Use the MATLAB program ODEBVP to solve the boundary value problem for $h = 0.1, 0.05, 0.025$, and print the errors of the numerical solutions at $t = 1.2$, $1.4, 1.6, 1.8$. Comment on how errors decrease when h is halved. Do the same for the extrapolated solutions.

8. The general solution of the equation

$$t^2 Y'' - t(t+2)Y' + (t+2)Y = 0$$

is $Y(t) = c_1 t + c_2 t e^t$. Determine the solution of the equation with the boundary conditions

$$Y(1) = e, \qquad Y(2) = 2e^2.$$

Use the MATLAB program ODEBVP to solve the boundary value problem for $h = 0.1, 0.05, 0.025$, print the errors of the numerical solutions at $t = 1.2, 1.4$, 1.6 and 1.8. Comment on how errors decrease when h is halved. Do the same for the extrapolated solutions.

9. The general solution of the equation

$$t Y'' - (2t+1)Y' + (t+1)Y = 0$$

is $Y(t) = c_1 e^t + c_2 t^2 e^t$. Find the solution of the equation with the boundary conditions

$$Y'(1) = 0, \qquad Y(2) = e^2.$$

Write down a formula for a discrete approximation of the boundary condition $Y'(1) = 0$ similar to (11.27), which has an accuracy $\mathcal{O}(h^2)$. Implement the method by modifying the program ODEBVP, and solve the problem with $h = 0.1$, $0.05, 0.025$. Print the errors of the numerical solutions at $t = 1, 1.2, 1.4, 1.6$,

1.8, and comment on how errors decrease when h is halved. Do the same for the extrapolated solutions.

10. Consider the boundary value problem (11.2) with p, q, and r constant. Modify the MATLAB program so that the command feval does not appear. Use the modified program to solve the following boundary value problem.

(a)
$$Y'' = -Y, \quad 0 < t < \tfrac{\pi}{2},$$
$$Y(0) = Y\left(\tfrac{1}{2}\pi\right) = 1.$$

The true solution is $Y(t) = \sin t + \cos t$.

(b)
$$Y'' + Y = \sin t, \quad 0 < t < \tfrac{\pi}{2},$$
$$Y(0) = Y\left(\tfrac{1}{2}\pi\right) = 0.$$

The true solution is $Y(t) = -\tfrac{1}{2}t\cos t$.

11. Give a second-order scheme for the following boundary value problem.

$$Y'' = \sin(tY') + 1, \quad 0 < t < 1,$$
$$Y(0) = 0, \quad Y(1) = 1.$$

12. Consider modifying the material of Section 11.1 to solve the BVP

$$Y''(t) = p(t)Y'(t) + q(t)Y(t) + r(t), \quad a < t < b,$$
$$Y(a) = g_1, \quad Y'(b) + kY(b) = g_2.$$

Do so with the first-order approximation given in (11.25). Give the analogs of the results (11.8)–(11.14).

13. Continuing with the preceding problem, modify ODEBVP to handle this new boundary condition. Apply it to the boundary value problem

$$\begin{cases} Y'' = -\dfrac{2t}{1+t^2}Y' + Y + \dfrac{2}{1+t^2} - \log(1+t^2), & 0 < t < 1, \\ Y(0) = 0, \quad Y'(1) + Y(1) = 1 + \log(2). \end{cases}$$

The true solution is $Y(t) = \log(1+t^2)$, just as with the earlier example (11.19). Repeat the calculations leading to Table 11.1. Check the assertion on the order of convergence given in Section 11.1.3 in the sentence containing (11.25).

14. Consider showing that the tridiagonal matrix T of (11.40) is nonsingular. For simplicity, denote its order by $m \times m$. To show that T is nonsingular, it is sufficient to show that the only solution $x \in \mathbb{R}^m$ of the homogeneous linear system $Tx = 0$ is the zero solution $x = 0$. Let $c = \max_{1 \le j \le m} |x_j|$. We want

to show $c = 0$. Begin by assuming the contrary, namely that $c > 0$. Write the individual equations in the system $Tx = 0$. In particular, consider an equation corresponding to a component of x that has magnitude c (of which there must be at least one), and denote its index by k. Assume initially that $1 < k < m$. Show from equation k that x_{k+1} and x_{k-1} must also have magnitude c. By induction, show that all components must have magnitude c; and then show from the first or last equation that this leads to a contradiction.

15. For each of the following BVPs for a second-order differential equation, consider converting it to an equivalent BVP for a system of first-order equations, as in (11.35). What are the matrices A and B of (11.35)?

   (a) The linear BVP (11.2)–(11.3).

   (b) The nonlinear BVP of (11.44).

   (c) The nonlinear BVP (11.29)–(11.30).

   (d) The following system of second-order equations: for $0 < t < 1$,

   $$mx''(t) = \frac{cx(t)}{(x(t)^2 + y(t)^2)^{3/2}}, \qquad my''(t) = \frac{cy(t)}{(x(t)^2 + y(t)^2)^{3/2}},$$

   with the boundary conditions

   $$\begin{aligned} x(0) &= x(1), & y(0) &= y(1), \\ x'(0) &= x'(1), & y'(0) &= y'(1). \end{aligned}$$

16. Consider converting nonzero boundary conditions to zero boundary conditions.

   (a) Consider the two-point boundary value problem (11.36). To convert this to an equivalent problem with zero boundary conditions, write $Y(x) = z(x) + w(x)$ with $w(x)$ a straight line satisfying the following boundary conditions: $w(a) = \gamma_1$, $w(b) = \gamma_2$. Derive a new boundary value problem for $z(x)$.

   (b) Generalize this procedure to problem (11.29). Obtain a new problem with zero boundary conditions. What assumptions, if any, are needed for the coefficients a_0, a_1, b_0, and b_1?

17. Using the shooting method of Subsection 11.2.2, solve the following boundary-value problems. Study the convergence rate as h is varied.

   (a) $Y'' = -\dfrac{2}{x}YY'$, $1 < x < 2$; $\quad Y(1) = \frac{1}{2}, Y(2) = \frac{2}{3}$.
   True solution: $Y(x) = x/(1 + x)$.

   (b) $Y'' = 2YY'$, $0 < x < \frac{1}{4}\pi$; $\quad Y(0) = 0, Y\left(\frac{1}{4}\pi\right) = 1$.
   True solution: $Y(x) = \tan(x)$.

CHAPTER 12

VOLTERRA INTEGRAL EQUATIONS

In earlier chapters the initial value problem

$$Y'(s) = f(s, Y(s)), \quad t_0 \le s \le b,$$

$$Y(t_0) = Y_0$$

was reformulated using integration. In particular, by integrating over the interval $[t_0, t]$, we obtain

$$Y(t) = Y_0 + \int_{t_0}^{t} f(s, Y(s)) \, ds, \quad t_0 \le t \le b.$$

This is an integral equation of Volterra type. Motivated in part by this reformulation, we consider now the integral equation

$$Y(t) = g(t) + \int_{0}^{t} K(t, s, Y(s)) \, ds, \quad 0 \le t \le T. \tag{12.1}$$

In this equation, the functions $K(t, s, u)$ and $g(t)$ are given; the function $Y(t)$ is unknown and is to be determined on the interval $0 \le t \le T$. This equation is called a *Volterra integral equation of the second kind*. Such integral equations occur in a variety of physical applications, and few of them can be reformulated easily as differential equation initial value problems. However, the numerical methods for such equations are linked to those for the initial value problem, and we consider such methods in this chapter.

12.1 SOLVABILITY THEORY

We begin by discussing some of the theory behind such equations, beginning with the linear equation

$$Y(t) = g(t) + \int_0^t K(t,s) Y(s) \, ds, \qquad 0 \le t \le T. \tag{12.2}$$

The function $K(t,s)$ is called the "kernel function" of the integral operator, or simply the "kernel". An important theoretical tool for studying this equation is the use of "successive approximations" or "Picard iteration".

As an initial estimate of the solution, choose $Y_0(t) \equiv g(t)$. Then define a sequence of iterates $\{Y_\ell(t)\}$ by

$$Y_{\ell+1}(t) = g(t) + \int_0^t K(t,s) Y_\ell(s) \, ds, \qquad 0 \le t \le T$$

for $\ell = 0, 1, \ldots$ To develop some intuition, we calculate $Y_2(t)$:

$$
\begin{aligned}
Y_2(t) &= g(t) + \int_0^t K(t,s) Y_1(s) \, ds \\
&= g(t) + \int_0^t K(t,s) \left[g(s) + \int_0^s K(s,v) g(v) \, dv \right] ds \\
&= g(t) + \int_0^t K(t,s) g(s) \, ds \\
&\quad + \int_0^t K(t,s) \int_0^s K(s,v) g(v) \, dv \, ds.
\end{aligned}
\tag{12.3}
$$

We then introduce a change in the order of integration,

$$
\begin{aligned}
\int_0^t \int_0^s K(t,s) \, K(s,v) \, g(v) \, dv \, ds \\
= \int_0^t g(v) \int_v^t K(t,s) \, K(s,v) \, ds \, dv.
\end{aligned}
\tag{12.4}
$$

and define

$$K_2(t,v) = \int_v^t K(t,s) K(s,v) \, ds, \qquad 0 \le v \le t \le T.$$

Then (12.3) becomes

$$Y_2(t) = g(t) + \int_0^t K(t,s)\,g(s)\,ds + \int_0^t K_2(t,v)\,g(v)\,dv.$$

This can be continued inductively to give

$$Y_\ell(t) = g(t) + \sum_{j=1}^{\ell} \int_0^t K_j(t,s)\,g(s)\,ds \tag{12.5}$$

for $\ell = 1, 2, \ldots$ The kernel functions K_j are defined by

$$K_1(t,s) = K(t,s),$$

$$K_j(t,s) = \int_s^t K(t,u)\,K_{j-1}(u,s)\,du, \qquad j = 2, 3, \ldots. \tag{12.6}$$

Much of the theory of solvability of the integral equation (12.2) can be developed by looking at the limit of (12.5) as $\ell \to \infty$. This, in turn, requires an examination of the kernel functions $\{K_j(t,s)\}_{j=1}^{\infty}$. Doing so yields the following theorem.

Theorem 12.1 *Assume that $K(t,s)$ is continuous for $0 \le s \le t \le T$, and that $g(t)$ is continuous on $[0,T]$. Then (12.2) has a unique continuous solution $Y(t)$ on $[0,T]$, and*

$$|Y(t)| \le e^{Bt} \max_{0 \le s \le t} |g(s)|, \tag{12.7}$$

where $B = \max_{0 \le s \le t \le T} |K(t,s)|$.

Some details of the proof are taken up in the problems.

A related approach can be used to prove the following theorem for the fully non-linear equation (12.1). The Picard iteration is now

$$Y_{\ell+1}(t) = g(t) + \int_0^t K(t,s,Y_\ell(s))\,ds, \qquad 0 \le t \le T$$

for $\ell = 0, 1, \ldots$

Theorem 12.2 *Assume that the function $K(t,s,u)$ satisfies the following two conditions:*
(a) $K(t,s,u)$ is continuous for $0 \le s \le t \le T$ and $-\infty < u < \infty$.
(b) $K(t,s,u)$ satisfies a Lipschitz condition,

$$|K(t,s,u_1) - K(t,s,u_2)| \le c\,|u_1 - u_2|, \qquad 0 \le s \le t \le T$$

for all $-\infty < u_1, u_2 < \infty$, with some $c > 0$.
Assume further that $g(t)$ is continuous on $[0,T]$. Then equation (12.1) has a unique continuous solution $Y(t)$ on the interval $[0,T]$. In addition,

$$|Y(t)| \le e^{ct} \max_{0 \le s \le t} |g(s)|. \tag{12.8}$$

For a proof, see Linz [59, Chap. 4].

As with differential equations, it is important to examine the stability of the solution $Y(t)$ with respect to changes in the data of the equation, K and g. We consider only the perturbation of the linear equation (12.2) by changing $g(t)$ to $g(t) + \varepsilon(t)$. Let $Y(t;\varepsilon)$ denote the solution of the perturbed equation,

$$Y(t;\varepsilon) = g(t) + \varepsilon(t) + \int_0^t K(t,s)\, Y(s;\varepsilon)\, ds, \qquad 0 \le t \le T. \tag{12.9}$$

Subtracting (12.2), we have

$$\begin{aligned} Y(t;\varepsilon) - Y(t) &= \varepsilon(t) \\ &+ \int_0^t K(t,s)\, [Y(s;\varepsilon) - Y(s)]\, ds, \qquad 0 \le t \le T. \end{aligned} \tag{12.10}$$

Applying (12.7) from Theorem 12.1, we have

$$|Y(t;\varepsilon) - Y(t)| \le e^{Bt} \max_{0 \le s \le t} |\varepsilon(s)|. \tag{12.11}$$

This shows stability of the solution with respect to perturbations in the function g in (12.2). This is a conservative estimate because the multiplying factor e^{Bt} increases very rapidly with t. The analysis of stability can be improved by examining (12.10) in greater detail, just as was done for differential equations in (1.16) of Section 1.2. We can also generalize these results to the nonlinear equation (12.1); see [59], [64].

12.1.1 Special equations

A model equation for studying the numerical solution of (12.1) is the simple linear equation

$$Y(t) = g(t) + \lambda \int_0^t Y(s)\, ds, \qquad t \ge 0. \tag{12.12}$$

This can be reformulated as the initial value problem

$$\begin{aligned} Y'(t) &= \lambda Y(t) + g'(t), \qquad t \ge 0, \\ Y(0) &= g(0), \end{aligned} \tag{12.13}$$

which is the model equation used in earlier chapters for studying numerical methods for solving the initial value problem for ordinary differential equations. Using the solution of this simple linear initial value problem leads to

$$Y(t) = g(t) + \lambda \int_0^t e^{\lambda(t-s)} g(s)\, ds, \qquad t \ge 0. \tag{12.14}$$

Recall from (1.20) of Section 1.2 that, usually, (12.13) is considered stable for $\lambda < 0$ and is considered unstable for $\lambda > 0$. Thus the same is true of the Volterra equation (12.12).

As another model Volterra integral equation, consider

$$Y(t) = g(t) + \lambda \int_0^t e^{\beta(t-s)} Y(s) \, ds, \qquad t \geq 0. \tag{12.15}$$

This can be reduced to the form of (12.12), and this leads to the solution

$$Y(t) = g(t) + \lambda \int_0^t e^{(\lambda+\beta)(t-s)} g(s) \, ds, \qquad t \geq 0. \tag{12.16}$$

Equations of the form

$$Y(t) = g(t) + \lambda \int_0^t K(t-s) Y(s) \, ds, \qquad t \geq 0 \tag{12.17}$$

are said to be of 'convolution type', and the *Laplace transform* can often be used to obtain a solution. Discussion of the Laplace transform and its application in solving differential equations can be found in most undergraduate textbooks on ordinary differential equations; for example, see [16]. Let $\widehat{K}(\tau)$ denote the Laplace transform of $K(t)$, and let $L(t; \lambda)$ denote the inverse Laplace transform of

$$\frac{\widehat{K}(\tau)}{1 - \lambda \widehat{K}(\tau)}.$$

The solution of (12.17) is given by

$$Y(t) = g(t) + \lambda \int_0^t L(t-s; \lambda) g(s) \, ds, \qquad t \geq 0. \tag{12.18}$$

Both (12.12) and (12.15) are special cases of (12.17).

12.2 NUMERICAL METHODS

Numerical methods for solving the Volterra integral equation

$$Y(t) = g(t) + \int_0^t K(t, s, Y(s)) \, ds, \qquad 0 \leq t \leq T \tag{12.19}$$

are similar to numerical methods for the initial value problem for ordinary differential equations. A set of grid points $\{t_i : i = 0, 1, \dots\}$ is chosen, and an approximation to $\{Y(t_i) : i = 0, 1, \dots\}$ is computed in a step-by-step procedure. For simplicity, we use an equally spaced grid,

$$t_i = ih, \qquad i = 0, 1, \dots, N_h,$$

where $hN_h \leq T$ and $h(N_h + 1) > T$. To aid in developing some intuition for this topic, we begin with an important special case, the *trapezoidal method*. Later a general scheme is given for the numerical approximation of (12.19). As with numerical methods for ordinary differential equations, let y_n denote an approximation of $Y(t_n)$. From (12.19), take $y_0 = Y(0) = g(0)$.

12.2.1 The trapezoidal method

For $n > 0$, write

$$Y(t_n) = g(t_n) + \int_0^{t_n} K(t_n, s, Y(s))\, ds.$$

Using the trapezoidal numerical integration rule, we obtain

$$\int_0^{t_n} K(t_n, s, Y(s))\, ds \approx h \sum_{j=0}^{n} {}'' K(t_n, t_j, Y(t_j)). \tag{12.20}$$

In this formula, the double-prime superscript indicates that the first and last terms should be halved before being summed. Using this approximation leads to the numerical formula

$$Y(t_n) \approx g(t_n) + h \sum_{j=0}^{n} {}'' K(t_n, t_j, Y(t_j)),$$

$$y_n = g(t_n) + h \sum_{j=0}^{n} {}'' K(t_n, t_j, y_j), \qquad n = 1, 2, \ldots, N_h. \tag{12.21}$$

This equation defines y_n implicitly, as earlier with the trapezoidal rule (4.22) of Section 4.2 for the initial value problem. Also, as before, when h is sufficiently small, this can be solved for y_n by simple fixed point iteration,

$$y_n^{(k+1)} = g(t) + \frac{h}{2} K(t_n, t_0, y_0)$$

$$+ h \sum_{j=1}^{n-1} K(t_n, t_j, y_j) + \frac{h}{2} K\left(t_n, t_n, y_n^{(k)}\right), \qquad k = 0, 1, \ldots \tag{12.22}$$

with some given $y_n^{(0)}$. Newton's method and other rootfinding methods can also be used. A MATLAB® program implementing (12.21)–(12.22) is given at the end of the section.

Example 12.3 Consider solving the equation

$$Y(t) = \cos t - \int_0^t Y(s)\, ds, \qquad t \geq 0 \tag{12.23}$$

with the true solution

$$Y(t) = \frac{1}{2}\left(\cos t - \sin t + e^{-t}\right), \qquad t \geq 0.$$

Equation (12.23) is the model equation (12.12) with $\lambda = -1$ and $g(t) = \cos t$. Numerical results for the use of (12.21) are shown in Table 12.1 for varying stepsizes h. It can be seen that the error at each value of t is of size $\mathcal{O}(h^2)$. ∎

Table 12.1 Numerical results for solving (12.23) using the trapezoidal method (12.21)

t	$h = 0.2$	Ratio	Error $h = 0.1$	Ratio	$h = 0.05$
0.8	$1.85e - 4$	4.03	$4.66e - 5$	4.01	$1.17e - 5$
1.6	$9.22e - 4$	4.03	$2.31e - 4$	4.01	$5.77e - 5$
2.4	$1.74e - 3$	4.03	$4.36e - 4$	4.01	$1.09e - 4$
3.2	$1.95e - 3$	4.03	$4.88e - 4$	4.01	$1.22e - 4$
4.0	$1.25e - 3$	4.04	$3.11e - 4$	4.01	$7.76e - 5$

12.2.2 Error for the trapezoidal method

To build some intuition for the behaviour of (12.21), we consider first the linear case (12.2),

$$y_n = g(t_n) + h \sum_{j=0}^{n}{}'' K(t_n, t_j) y_j, \qquad n = 1, 2, \ldots, N_h. \tag{12.24}$$

Rewrite the original equation (12.2) using the trapezoidal numerical integration rule with its error formula,

$$Y(t_n) = g(t_n) + h \sum_{j=0}^{n}{}'' K(t_n, t_j) Y(t_j) + Q_h(t_n), \tag{12.25}$$

for $n = 1, 2, \ldots, N_h$. The error term can be written in various forms:

$$Q_h(t_n) = -\sum_{j=1}^{n} \frac{h^3}{12} \frac{\partial^2}{\partial s^2} [K(t_n, s) Y(s)] \Big|_{s=\tau_{n,j}} \tag{12.26}$$

$$= -\frac{h^2 t_n}{12} \frac{\partial^2}{\partial s^2} [K(t_n, s) Y(s)] \Big|_{s=\tau_n} \tag{12.27}$$

$$\approx -\frac{h^2}{12} \frac{\partial}{\partial s} [K(t_n, s) Y(s)] \Big|_{s=0}^{t_n}. \tag{12.28}$$

In (12.26), $\tau_{n,j}$ is some unknown point in $[t_{j-1}, t_j]$; and in (12.27), τ_n is an unknown point in $[0, t_n]$. These are standard error formulas for the trapezoidal quadrature rule; e.g. see [12, §5.2]. Subtract (12.24) from (12.25), obtaining

$$E_h(t_n) = h \sum_{j=0}^{n}{}'' K(t_n, t_j) E_h(t_j) + Q_h(t_n) \tag{12.29}$$

in which $E_h(t_n) = Y(t_n) - y_n$.

Example 12.4 As a simple particular case of (12.24), choose $K(t, s) \equiv \lambda$ and $Y(s) = s^2$. We are solving the equation (12.12) with a suitable choice of $g(t)$.

Using (12.27) and noting that $E_h(t_0) = E_h(0) = 0$, (12.29) becomes

$$E_h(t_n) = \sum_{j=1}^{n-1} h\lambda E_h(t_j) + \frac{h\lambda}{2} E_h(t_n) - \frac{h^2 t_n}{12} Y''(\tau_n).$$

Because $Y''(s) \equiv 2$, this simplifies further to

$$E_h(t_n) = \sum_{j=1}^{n-1} h\lambda E_h(t_j) + \tfrac{1}{2} h\lambda E_h(t_n) - \tfrac{1}{6} h^2 t_n, \qquad (12.30)$$

for $n = 1, \ldots, N_h$. This complicated expression can be solved explicitly.

Write the same formula with $n - 1$ replacing n, and then subtract it from (12.30). This yields

$$E_h(t_n) - E_h(t_{n-1}) = h\lambda E_h(t_{n-1}) + \tfrac{1}{2} h\lambda E_h(t_n) - \tfrac{1}{2} h\lambda E_h(t_{n-1})$$
$$- \tfrac{1}{6} h^2 (t_n - t_{n-1}).$$

Solving for $E_h(t_n)$, we obtain

$$E_h(t_n) = \left(\frac{1 + \tfrac{1}{2} h\lambda}{1 - \tfrac{1}{2} h\lambda}\right) E_h(t_{n-1}) - \frac{1}{1 - \tfrac{1}{2} h\lambda} \frac{h^3}{6}, \qquad n \geq 0.$$

Using induction, this has the solution

$$E_h(t_n) = \left(\frac{1 + \tfrac{1}{2} h\lambda}{1 - \tfrac{1}{2} h\lambda}\right)^n E_h(t_0) - \left[\sum_{j=0}^{n-1} \left(\frac{1 + \tfrac{1}{2} h\lambda}{1 - \tfrac{1}{2} h\lambda}\right)^j\right] \frac{1}{1 - \tfrac{1}{2} h\lambda} \frac{h^3}{6}. \quad (12.31)$$

The first term equals zero since $E_h(t_0) = 0$; and the second term involves a geometric series which sums to

$$\frac{\left(\dfrac{1 + \tfrac{1}{2} h\lambda}{1 - \tfrac{1}{2} h\lambda}\right)^n - 1}{\left(\dfrac{1 + \tfrac{1}{2} h\lambda}{1 - \tfrac{1}{2} h\lambda}\right) - 1} = \frac{2 - h\lambda}{2h\lambda} \left\{\left[1 + \frac{h\lambda}{1 - \tfrac{1}{2} h\lambda}\right]^n - 1\right\}.$$

Using this in (12.31),

$$E_h(t_n) = -\frac{h^2}{6\lambda} \left\{\left[1 + \frac{h\lambda}{1 - \tfrac{1}{2} h\lambda}\right]^n - 1\right\}.$$

For a fixed $t = t_n = nh$, as $h \to 0$, this can be manipulated to obtain the asymptotic formula

$$E_h(t_n) \approx -\frac{h^2}{6\lambda} \left(e^{\lambda t_n} - 1\right). \qquad \blacksquare$$

For this special case, the numerical solution of (12.12) using the trapezoidal method has an error of size $\mathcal{O}(h^2)$. This is of the same order in h as the discretization error for the trapezoidal rule approximation in (12.20). Although this result has been shown for only a special solution, it turns out to be true in general for the trapezoidal method of (12.21). This is discussed in greater detail in Section 12.3, including a general convergence theorem that includes the trapezoidal rule being applied to the fully nonlinear equation (12.19).

12.2.3 General schema for numerical methods

As a general approach to the numerical solution of the integral equation (12.19), consider replacing the integral term with an approximation based on numerical integration. Introduce the numerical integration

$$\int_0^{t_n} K(t_n, s, Y(s))\ ds \approx h \sum_{j=0}^n w_{n,j} K(t_n, t_j, Y(t_j)). \tag{12.32}$$

The quadrature weights $hw_{n,j}$ are allowed to vary with the grid point t_n, in contrast to the trapezoidal method. Equation (12.19) is approximated by

$$y_n = g(t_n) + h \sum_{j=0}^n w_{n,j} K(t_n, t_j, y_j), \qquad n = 1, 2, \ldots, N_h. \tag{12.33}$$

As with the earlier trapezoidal method, if $w_{n,n} \neq 0$, then (12.33) must be solved for y_n by some rootfinding method. For example, simple iteration has the form

$$\begin{aligned} y_n^{(k+1)} = g(t_n) + h \sum_{j=0}^{n-1} w_{n,j} K(t_n, t_j, y_j) \\ + h w_{n,n} K\left(t_n, t_n, y_n^{(k)}\right), \qquad k = 0, 1, \ldots \end{aligned} \tag{12.34}$$

for some given initial estimate $y_n^{(0)}$. Also, many such methods (12.33) require $n \geq p + 1$ for some small integer p; the values $y_1, \ldots, y_p$ must be determined by some other "starting method".

There are many possible such schemes (12.33), and we investigate only one pair of such formulas, both based on Simpson's numerical integration formula. The simple Simpson rule has the form

$$\int_\alpha^{\alpha+2h} F(s)\ ds \approx \frac{h}{3} \left[F(\alpha) + 4F(\alpha + h) + F(\alpha + 2h) \right].$$

This classical quadrature formula is very popular, well-studied, and well-understood; e.g., see [12, Sections 5.1–5.2]. In producing the approximation of (12.32), consider first the case where n is even. Then define

$$
\int_0^{t_n} K(t_n, s, Y(s))\, ds = \sum_{j=1}^{n/2} \int_{t_{2j-2}}^{t_{2j}} K(t_n, s, Y(s))\, ds
$$

$$
\approx \frac{h}{3} \sum_{j=1}^{n/2} \big[K(t_n, t_{2j-2}, Y(t_{2j-2})) + 4K(t_n, t_{2j-1}, Y(t_{2j-1}))
$$

$$
+ K(t_n, t_{2j}, Y(t_{2j})) \big] \ . \tag{12.35}
$$

This has an error of size $\mathcal{O}(h^4)$.

Consider next the case that n where odd and $n \geq 3$. Then the interval $[0, t_n]$ cannot be divided into a union of subintervals $[t_{2j-2}, t_{2j}]$; and thus Simpson's integration rule cannot be applied in the manner of (12.35). To maintain the accuracy implicit in using Simpson's rule, we use Newton's $\frac{3}{8}$'s rule over one subinterval of length $3h$,

$$
\int_\alpha^{\alpha + 3h} F(s)\, ds \approx \frac{3h}{8} \left[F(\alpha) + 3F(\alpha + h) + 3F(\alpha + 2h) + F(a + 3h) \right].
$$

We then use Simpson's rule over the remaining subintervals of length $2h$. The interval $[0, t_n]$ can be subdivided in two convenient ways,

Scheme 1: $[0, t_n] = [0, t_3] \cup [t_3, t_5] \cup \cdots \cup [t_{n-2}, t_n]$; (12.36)

Scheme 2: $[0, t_n] = [0, t_2] \cup \cdots \cup [t_{n-5}, t_{n-3}] \cup [t_{n-3}, t_n]$. (12.37)

With the first scheme, we apply Newton's $\frac{3}{8}$'s rule over $[0, t_3]$ and apply Simpson's rule over the subintervals $[t_3, t_5], \ldots, [t_{n-2}, t_n]$. With the second scheme, we apply Newton's $\frac{3}{8}$'s rule over $[t_{n-3}, t_n]$ and Simpson's rule over the remaining subintervals $[0, t_2], \ldots, [t_{n-5}, t_{n-3}]$.

To be more precise, with the second scheme we begin by writing

$$
\int_0^{t_n} K(t_n, s, Y(s))\, ds = \sum_{j=1}^{(n-3)/2} \int_{t_{2j-2}}^{t_{2j}} K(t_n, s, Y(s))\, ds
$$

$$
+ \int_{t_{n-3}}^{t_n} K(t_n, s, Y(s))\, ds.
$$

Approximating the integrals as described above, we obtain

$$
\int_0^{t_n} K(t_n, s, Y(s))\, ds \approx \frac{1}{3} h \sum_{j=1}^{n/2} \{ K(t_n, t_{2j-2}, Y(t_{2j-2}))
$$

$$
+ 4K(t_n, t_{2j-1}, Y(t_{2j-1})) + K(t_n, t_{2j}, Y(t_{2j}))] \tag{12.38}
$$

$$
+ \frac{3}{8} h \{ K(t_n, t_{n-3}, Y(t_{n-3})) + 3K(t_n, t_{n-2}, Y(t_{n-2}))
$$

$$
+ 3K(t_n, t_{n-1}, Y(t_{n-1})) + K(t_n, t_n, Y(t_n)) \} .
$$

Using (12.36) leads to a similar formula, but with Newton's $\frac{3}{8}$'s rule applied over $[0, t_3]$.

We denote by "Simpson method 2" the combination of (12.35) and (12.38); and we denote as "Simpson method 1" the combination of (12.35) and the analog of (12.38) for the subdivision of (12.36). Both methods require that the initial value y_1 be calculated by another method.

Both approximations have discretization errors of size $\mathcal{O}(h^4)$, but method 2 turns out to be much superior to method 1 when solving (12.19). These methods are discussed and illustrated in Section 12.3.

MATLAB program. The following MATLAB program implements the trapezoidal method (12.21)–(12.22).

```
function soln = vie_trap(N_h,T,fcn_g,fcn_k)
%
% function soln = vie_trap(N_h,T,fcn_g,fcn_k)
%
% This solves the integral equation
%                t
% Y(t) = g(t) + Int k(t,s,Y(s))ds
%                0
% ==INPUT==
% N_h:    The number of subdivisions of [0,T].
% T:      [0,T] is the interval for the solution function.
% fcn_g:  The handle of the driver function g(t).
% fcn_k:  The handle of the kernel function k(t,s,u).

% ==OUTPUT==
% soln:    A structure with the following components.
% soln.t:  The grid points at which the solution Y(t) is
%          approximated.
% soln.y:  The approximation of Y(t) at the grid points.

% The implicit trapezoidal equation is solved by simple fixed
% point iteration at each grid point in t.  For simplicity,
% the program uses a crude means of controlling the iteration.
% The iteration is executed a fixed number of times, controlled
% by 'loop'.
loop = 10; % This is much more than is usually needed.

h = T/N_h; t = linspace(0,T,N_h+1);
g_vec = fcn_g(t);
g_vec = zeros(size(t)); y_vec(1) = g_vec(1);
for n=1:N_h
  y_vec(n+1) = y_vec(n); % Initial estimate for the iteration.
  k_vec = fcn_k(t(n+1),t(1:n+1),y_vec(1:n+1));
```

```
   for j=1:loop
     y_vec(n+1) = g_vec(n+1) + h*(sum(k_vec(2:n)) ...
                    + (k_vec(1) + k_vec(n+1))/2);
     k_vec(n+1) = fcn_k(t(n+1),t(n+1),y_vec(n+1));
   end
 end
 soln.t = t;
 soln.y = y_vec;
 end % vie_trap
```

The following program is a test program for the above vie_trap.

```
function test_vie_trap(lambda,N_h,T,output_step)
%
% function test_vie_trap(lambda,N_h,T,output_step)
%
% ==INPUT==
% lambda:   Used in defining the integral equation.
% N_h:      The number of subdivisions of [0,T].
% T:        [0,T] is the interval for the solution function.
% output_step: The solution is output at the indices
%             v = 1:output_step:N_h+1

soln = vie_trap(N_h,T,@g_driver,@kernel);
t = soln.t; y = soln.y;
true = true_soln(t);
error = true - y;
format short e
v = 1:output_step:N_h+1;
disp([t(v)' y(v)' error(v)'])

%=================================================
function ans_g = g_driver(s)
ans_g = (1-lambda)*sin(t) + (1+lambda)*cos(t) - lambda;
end % g_driver

function ans_true = true_soln(s)
ans_true = cos(s) + sin(s);
end % true_soln

function ans_k = kernel(tau,s,u)
% tau is a scalar, s and u vectors of the same dimension.
ans_k = lambda*u;
end % kernel
%=================================================
end % test_vie_trap
```

12.3 NUMERICAL METHODS: THEORY

We begin by considering the convergence of methods

$$y_n = g(t_n) + h \sum_{j=0}^{n} w_{n,j} K(t_n, t_j, y_j), \qquad n = p+1, \ldots, N_h \qquad (12.39)$$

with $y_0 = g(0)$ and with $y_1, \ldots, y_p$ determined by another method. For example, the trapezoidal method has $p = 0$, and the two Simpson methods discussed in and following (12.35) have $p = 1$. Later we discuss the error requirements when computing such initial values $y_1, \ldots, y_p$.

To analyze the error in using (12.39) to solve

$$Y(t) = g(t) + \int_0^t K(t, s, Y(s))\, ds, \qquad 0 \le t \le T, \qquad (12.40)$$

we proceed in analogy with the error equation (12.29) for the trapezoidal method. As in Section 12.1, we assume that $K(t, s, u)$ is continuous for $0 \le s \le t \le T$; further, we assume that $K(t, s, u)$ satisfies the Lipschitz condition

$$|K(t, s, u_1) - K(t, s, u_2)| \le c |u_1 - u_2|, \qquad 0 \le s \le t \le T \qquad (12.41)$$

for $-\infty < u_1, u_2 < \infty$. These are the assumptions used in Theorem 12.2.

Rewrite (12.40) using numerical integration and the associated error,

$$Y(t_n) = g(t_n) + h \sum_{j=0}^{n} w_{n,j} K(t_n, t_j, Y(t_j))$$

$$+ Q_h(t_n), \qquad n = p+1, \ldots, N_h. \qquad (12.42)$$

The quantity $Q_h(t_n)$ denotes the error in the quadrature approximation to the integral in (12.40). As an example of the quadrature error, recall (12.25)–(12.28) for the trapezoidal method.

Subtract (12.39) from (12.42), obtaining

$$E_h(t_n) = h \sum_{j=0}^{n} w_{n,j} \left[K(t_n, t_j, Y(t_j)) - K(t_n, t_j, y_j) \right] + Q_h(t_n) \qquad (12.43)$$

for $n = p+1, \ldots, N_h$, with $E_h(t_n) = Y(t_n) - y_n$. Applying the Lipschitz condition (12.41) to (12.43), we have

$$|E_h(t_n)| \le hc \sum_{j=0}^{n} |w_{n,j}| |E_h(t_j)| + Q_h(t_n), \qquad n = p+1, \ldots, N_h. \qquad (12.44)$$

If we assume that h is small enough that $hc|w_{n,n}| < 1$, then we can bound $|E_h(t_n)|$ in terms of preceding errors:

$$|E_h(t_n)| \le \frac{hc}{1 - hc|w_{n,n}|} \sum_{j=0}^{n-1} |w_{n,j}| |E_h(t_j)| + \frac{Q_h(t_n)}{1 - hc|w_{n,n}|}, \qquad (12.45)$$

for $n = p + 1, \ldots, N_h$.

To further simplify this, we assume

$$\max_{0 \leq i \leq n \leq N_n} |w_{n,i}| \leq \gamma < \infty \tag{12.46}$$

for all $0 < h \leq h_0$ for some small value of h_0. Without any loss of generality when analyzing convergence as $h \to 0$, (12.46) permits the assumption that

$$hc\,|w_{n,n}| \leq \tfrac{1}{2} \tag{12.47}$$

is true for all h and n of interest. With (12.46) and (12.47), the inequality (12.45) becomes

$$|E_h(t_n)| \leq 2\gamma ch \sum_{j=0}^{n-1} |E_h(t_j)| + 2Q_h(t_n), \qquad n = p + 1, \ldots, N_h. \tag{12.48}$$

This can be solved to give a useful convergence result.

Theorem 12.5 *In the Volterra integral equation (12.40), assume that the function $K(t, s, u)$ is continuous for $0 \leq s \leq t \leq T$, $-\infty < u < \infty$, and further that it satisfies the Lipschitz condition (12.41). Assume that $g(t)$ is continuous on $[0, T]$. In the numerical approximation (12.39), assume (12.46). Introduce*

$$\eta(h) \equiv \sum_{j=0}^{p} |E_h(t_j)|, \tag{12.49}$$

$$\delta(t_n; h) \equiv \max_{p+1 \leq j \leq n} |Q_h(t_j)|.$$

Then

$$|E_h(t_n)| \leq e^{2\gamma ct_n} \left[2\gamma ch\eta(h) + \delta(t_n; h) \right], \qquad n = p + 1, \ldots, N_h. \tag{12.50}$$

Proof. This bound is a consequence of (12.48), the following lemma, and the bound

$$(1 + 2\gamma ch)^{n-p-1} \leq e^{2\gamma c(t_n - t_{p+1})} \leq e^{2\gamma ct_n}, \qquad n \geq p + 1.$$

To show this bound, recall Lemma 2.3 from Section 2.2. A more complete proof is given in [59, Section 7.3]. ∎

Lemma 12.6 *Let the sequence $\{\varepsilon_0, \varepsilon_1, \ldots\}$ satisfy*

$$|\varepsilon_n| \leq \alpha \sum_{j=0}^{n-1} |\varepsilon_j| + \beta_n, \qquad n = p + 1, \ldots. \tag{12.51}$$

Then

$$|\varepsilon_n| \leq (1 + \alpha)^{n-p-1} \left(\alpha \sum_{j=0}^{p} |\varepsilon_j| + \max_{p+1 \leq j \leq n} |\beta_j| \right). \tag{12.52}$$

Proof. This can be proved using mathematical inductions, and we leave it as an exercise for the reader. ■

The bound (12.50) assures us of convergence provided $h\eta(h) \to 0$ and $\delta(t_n; h) \to 0$ as $h \to 0$.

Example 12.7 Recall the trapezoidal method of (12.21). Then $p = 0$ and $\eta(h) = |Y(0) - y_0|$. For the purpose of analyzing convergence, we take $y_0 = Y(0)$ and $\eta(h) = 0$. Also, from (12.27), we can take

$$\delta(t_n; h) = -\frac{h^2 t_n}{12} \max_{0 \le s \le t_n} \left| \frac{\partial^2}{\partial s^2} [K(t_n, s) Y(s)] \right|. \tag{12.53}$$

From (12.50), we obtain

$$|E_h(t_n)| \le e^{2\gamma c t_n} \delta(t_n; h),$$

and this is of size $\mathcal{O}(h^2)$ on each finite interval $0 \le t_n \le T$. Thus the trapezoidal method is convergent; and we say it is of order 2.

Example 12.8 Recall Simpson method 2 from (12.35), (12.38), and the associated Simpson method 1. Both methods require $p = 1$, and

$$\eta(h) = |E_h(t_0)| + |E_h(t_1)|.$$

Again, we take $|E_h(t_0)| = 0$. The quadrature error $\delta(t_n; h)$ can be shown to be of size $\mathcal{O}(h^4)$ on each finite interval $[0, t_n]$. If we also have $h\eta(h) = \mathcal{O}(h^4)$, then the overall error in both Simpson methods is of size $\mathcal{O}(h^4)$ on each finite interval $[0, T]$.

If we use the simple trapezoidal method to generate y_1, then it can be shown that $\eta(h) = \mathcal{O}(h^3)$ for this special case of a fixed finite number of errors (in particular, $E_h(t_1)$); this is sufficient to yield $h\eta(h) = \mathcal{O}(h^4)$. We illustrate this using Simpson method 2 to solve

$$Y(t) = \cos t - \int_0^t Y(s)\, ds, \quad t \ge 0 \tag{12.54}$$

with the true solution

$$Y(t) = \tfrac{1}{2}\left(\cos t - \sin t + e^{-t}\right), \quad t \ge 0, \tag{12.55}$$

the same test equation as in example 12.3. The numerical results with varying values of h are given in Table 12.2. The values in the columns labeled *"Ratio" approach 16 as h decreases, and this is consistent with a convergence rate of $\mathcal{O}(h^4)$.* ■

12.3.1 Numerical stability

In addition to being convergent, a numerical method must also be numerically stable. As with numerical methods for the initial value problem for differential equations,

Table 12.2 Numerical results for solving (12.54) using the Simpson method 2

t	$h = 0.2$	Ratio	$h = 0.1$	Error Ratio	$h = 0.05$	Ratio	$h = 0.025$
0.8	$1.24e-6$	10.2	$1.23e-7$	13.4	$9.15e-9$	14.8	$6.16e-10$
1.6	$-5.56e-7$	-71.0	$7.84e-9$	6.4	$1.23e-9$	13.5	$3.09e-11$
2.4	$-1.90e-6$	14.2	$-1.34e-7$	14.3	$-9.37e-9$	15.1	$-6.22e-10$
3.2	$-1.95e-6$	10.4	$-1.87e-7$	13.6	$-1.38e-8$	14.9	$-9.24e-10$
4.0	$-7.10e-7$	6.2	$-1.15e-7$	12.9	$-8.95e-9$	14.7	$-6.07e-10$

various meanings are given to the concept of "numerically stable". We begin with stability as discussed in (12.9)-(12.11) for the linear equation (12.2). This is in analogy with stability as discussed in Section 7.3 of Chapter 7 for multistep methods for the initial value problem for differential equations.

In the numerical method

$$y_n = g(t_n) + h \sum_{j=0}^{n} w_{n,j} K(t_n, t_j, y_j), \quad n = p+1, \ldots, N_h. \quad (12.56)$$

consider perturbing the initial values $y_0, \ldots, y_p$, say, by changing them to $y_j + \eta_{h,j}$, $j = 0, \ldots, p$. Also, perturb $g(t_n)$ to $g(t_n) + \varepsilon_{h,n}$ for $n \geq p+1$. We are interested in knowing how the perturbations $\{\eta_{h,j}\}$ and $\{\varepsilon_{h,n}\}$ affect the solution $\{y_n\}$, particularly for small perturbations and small values of h.

Let $\{\tilde{y}_n : 0 \leq n \leq N_h\}$ denote the numerical solution in this perturbed case,

$$\tilde{y}_n = g(t_n) + \varepsilon_{h,n} + h \sum_{j=0}^{n} w_{n,j} K(t_n, t_j, \tilde{y}_j), \quad n = p+1, \ldots, N_h,$$

$$\quad (12.57)$$

$$\tilde{y}_n = y_n + \eta_{h,j}, \quad j = 0, \ldots, p.$$

Subtracting (12.56) from (12.57), using the Lipschitz condition (12.41) and the bound (12.46) for the weights, we obtain

$$|\tilde{y}_n - y_n| \leq |\varepsilon_{h,n}| + hc\gamma \sum_{j=0}^{n} |\tilde{y}_j - y_j|, \quad p+1 \leq n \leq N_h,$$

$$\tilde{y}_n - y_n = \eta_{h,j}, \quad j = 0, \ldots, p.$$

With assumption (12.47) and Lemma 12.6, we obtain

$$|\tilde{y}_n - y_n| \leq e^{2\gamma c t_n} \left(2h\gamma c \sum_{j=0}^{p} |\eta_{h,j}| + \max_{p+1 \leq j \leq n} |\varepsilon_{h,j}| \right).$$

This simplifies as

$$|\tilde{y}_n - y_n| \leq C\delta, \quad p+1 \leq n \leq N_h, \quad 0 < h \leq h_0, \quad (12.58)$$

where C is a constant independent of h and

$$\delta = \max_{0 < h \leq h_0} \left\{ h \max_{0 \leq j \leq p} |\eta_{h,j}|, \max_{p+1 \leq j \leq N_h} |\varepsilon_{h,j}| \right\}.$$

The upper bound h_0 on h is to be chosen so that for all n,

$$h_0 c \, |w_{n,n}| \leq \tfrac{1}{2}.$$

The bound (12.58) says that the numerical solution $\{y_n : p + 1 \leq n \leq N_h\}$ varies continuously with the initial starting values $\{y_0, \ldots, y_p\}$ and the function $g(t)$. This is true in a uniform sense for all sufficiently small values of h. The bound (12.58) is the numerical analogue of the stability result (12.11) for the linear equation (12.2).

The result (12.58) says that virtually all convergent quadrature schemes lead to numerical methods (12.56) that are numerically stable. In practice, however, a number of such methods remain very sensitive to perturbations in the starting values. In particular, experimental results imply that Simpson method 2 is numerically stable, whereas Simpson method 1 has practical stability problems. What is the explanation for this?

12.3.2 Practical numerical stability

In discussing practical stability difficulties when using numerical methods (12.39), we follow Linz [59, §7.4]. We consider only the linear equation

$$Y(t) = g(t) + \int_0^t K(t, s) \, Y(s) \, ds, \qquad 0 \leq t \leq T, \qquad (12.59)$$

although the results generalize to the fully nonlinear equation (12.40). The type of stability that is considered is related to the concept of "relative stability" from Subsection 7.3.3.

Consider the numerical method (12.39) as applied to (12.59),

$$y_n = g(t_n) + h \sum_{j=0}^{n} w_{n,j} K(t_n, t_j) \, y_j, \qquad n = p + 1, \ldots, N_h \qquad (12.60)$$

with $y_0 = g(0)$ and with $y_1, \ldots, y_p$ obtained by other means. The true solution $Y(t)$ satisfies

$$Y(t_n) = g(t_n) + h \sum_{j=0}^{n} w_{n,j} K(t_n, t_j) \, Y(t_j) + Q_h(t_n), \qquad (12.61)$$

for $n = p + 1, \ldots, N_h$. Subtracting (12.60) from (12.61), we obtain

$$E_h(t_n) = h \sum_{j=0}^{n} w_{n,j} K(t_n, t_j) \, E_h(t_j) + Q_h(t_n), \qquad (12.62)$$

for $n = p + 1, \ldots, N_h$.

To aid in understanding the behavior of $E_h(t_n)$ as t_n increases, the error is decomposed into two parts. First, let $\left\{ E_h^Q(t_n) \right\}$ denote the solution of

$$E_h^Q(t_n) = h \sum_{j=0}^{n} w_{n,j} K(t_n, t_j) E_h^Q(t_j) + Q_h(t_n), \qquad n = p + 1, \ldots, N_h,$$

$$E_h^Q(t_j) = 0, \qquad j = 0, \ldots, p.$$

(12.63)

This error is due entirely to the quadrature errors $\{Q_h(t_n) : n \geq p + 1\}$ that occur in discretizing the integral equation (12.59); it assumes that there is no error in the initial values $y_0, \ldots, y_p$. Second, consider the errors $E_h^S(t_n)$ obtained by solving

$$E_h^S(t_n) = h \sum_{j=0}^{n} w_{n,j} K(t_n, t_j) E_h^S(t_j), \qquad n = p + 1, \ldots, N_h,$$

(12.64)

$$E_h^S(t_j) = \eta_j, \qquad j = 0, \ldots, p.$$

(12.65)

The quantities $\{\eta_0, \ldots, \eta_p\}$ are the errors in the starting values $\{y_0, \ldots, y_p\}$ when using (12.60). The original error $E_h(t_n)$ is given by

$$E_h(t_n) = E_h^Q(t_n) + E_h^S(t_n), \qquad n = 0, 1, \ldots, N_h.$$

Returning to (12.63), assume that the quadrature error has an expansion of the form

$$Q_h(t_n) = a(t_n) h^m + \mathcal{O}(h^{m+1})$$

for some integer $m \geq 1$. For example, the trapezoidal method has

$$Q_h(t_n) = a(t) h^2 + \mathcal{O}(h^3),$$

$$a(t) = -\frac{1}{12} \frac{\partial}{\partial s} [K(t, s) Y(s)] \Big|_{s=0}^{t}$$

(see (12.28)). Then it can be shown that $E_h^Q(t_n)$ has the asymptotic formula

$$E_h^Q(t_n) = b(t_n) h^m + \mathcal{O}(h^{m+1})$$

(12.66)

with the function b the solution of the integral equation

$$b(t) = -a(t) + \int_0^t K(t, s) b(s) \, ds, \qquad 0 \leq t \leq T.$$

For a derivation of this, see [59, Theorem 7.3]. The asymptotic formula (12.66) applies to virtually all quadrature schemes that are likely to be used in setting up the numerical scheme (12.56), and it forms the basis for numerical extrapolation schemes for error estimation.

The second error, $E_h^S(t_n)$, is more subtle to understand. To begin, consider the weights $\{w_{n,j}\}$ for the two Simpson methods.

- Simpson method 1:

$$n \text{ even:} \quad \tfrac{1}{3}, \tfrac{4}{3}, \tfrac{2}{3}, \tfrac{4}{3}, \cdots, \tfrac{2}{3}, \tfrac{4}{3}, \tfrac{1}{3};$$

$$n \text{ odd:} \quad \tfrac{3}{8}, \tfrac{9}{8}, \tfrac{9}{8}, \tfrac{3}{8} + \tfrac{1}{3}, \tfrac{4}{3}, \tfrac{2}{3}, \tfrac{4}{3}, \cdots, \tfrac{2}{3}, \tfrac{4}{3}, \tfrac{1}{3}.$$

$$(12.67)$$

all being multiplied by h. The weights satisfy

$$w_{n+\rho,i} = w_{n,i}, \quad i = 4, \ldots, n$$

with $\rho = 2$, but not with $\rho = 1$. We say the weights have a *repetition factor* of 2.

- Simpson method 2:

$$n \text{ even:} \quad \tfrac{1}{3}, \tfrac{4}{3}, \tfrac{2}{3}, \tfrac{4}{3}, \cdots, \tfrac{2}{3}, \tfrac{4}{3}, \tfrac{1}{3};$$

$$n \text{ odd:} \quad \tfrac{1}{3}, \tfrac{4}{3}, \tfrac{2}{3}, \tfrac{4}{3}, \cdots, \tfrac{2}{3}, \tfrac{4}{3}, \tfrac{1}{3} + \tfrac{3}{8}, \tfrac{9}{8}, \tfrac{9}{8}, \tfrac{3}{8}.$$

$$(12.68)$$

The weights satisfy

$$w_{n+1,i} = w_{n,i}, \quad i = 0, 1, \ldots, n - 4.$$

and again, all being multiplied by h. These weights have a repetition factor of 1.

Both of these methods have an asymptotic formula for $E_h^S(t_n)$; see [59, Theorem 7.4].

In particular, for Simpson method 2 assume that the starting values $\{y_0, y_1\}$ satisfy

$$Y(t_i) - y_i = \delta_i h^3 + \mathcal{O}(h^4).$$

$$(12.69)$$

Then

$$E_h^S(t_n) = h^4 \left[\delta_0 C_0(t_n) + \delta_1 C_1(t_n) \right] + \mathcal{O}(h^5)$$

$$(12.70)$$

with $C_i(t)$ satisfying

$$C_i(t) = V_i K(t, t_i) + \int_0^t K(t, s) C_i(s) \, ds, \quad i = 0, 1, 2.$$

The constants V_i are derived as a part of the proof in [59, Theorem 7.4]. The functions $C_0(t)$ and $C_1(t)$ can be shown to be well behaved, and consequently, the same is true of the error in (12.70).

For Simpson method 1, there is an asymptotic formula for $E_h^S(t_n)$, but it is not as well behaved as is (12.70) for Simpson method 2. For Simpson method 1, it can be shown that

$$E_h^S(t_{2n}) = hx(t_{2n}) + \mathcal{O}(h^2),$$

$$(12.71)$$

$$E_h^S(t_{2n+1}) = hy(t_{2n+1}) + \mathcal{O}(h^2)$$

$$(12.72)$$

with $(x(t), y(t))$ the solution of a system of two Volterra integral equations. The functions $x(t)$ and $y(t)$ can be written in the form

$$x(t) = \tfrac{1}{2} \left(z_1(t) + z_2(t) \right),$$
$$y(t) = \tfrac{1}{2} \left(z_1(t) - z_2(t) \right)$$

(12.73)

with $z_1(t)$ and $z_2(t)$ the solutions of the Volterra integral equation

$$z_i(t) = g_i(t) + \int_0^t K(t, s)\, z_i(s)\, ds, \qquad 0 \le t \le T$$

for particular values of $g_i(t)$ that depend on both $K(t, s)$ and the constants $\{\delta_0, \delta_1\}$ of (12.69).

To develop some intuition from this, consider the special case $K(t, s) \equiv \lambda$. Then $z_1(t)$ and $z_2(t)$ have the forms

$$z_1(t) = A_1(t) + B_1(t)e^{\lambda t},$$
$$z_2(t) = A_2(t) + B_2(t)e^{-\lambda t/3}.$$

Recalling the special formulas of (12.12)–(12.14), the case $\lambda < 0$ is associated with stability in the Volterra integral equation and $\lambda > 0$ is associated with instability.

Considering only the case where $\lambda < 0$, the function $z_1(t)$ behaves "properly" as t increases. In contrast, the function $z_2(t)$ is exponentially increasing as t increases. Applying this to (12.73), we have that $x(t)$ and $y(t)$ will also increase exponentially, although with opposite signs depending on whether the index for t_n is even or odd. Using this in (12.71)-(12.72), we find that the errors $E_h^S(t_n)$ should increase exponentially for larger values of n, and that there should be an oscillation in sign.

Example 12.9 Recall Example 12.8 in which we examined Simpson method 2 for the linear integral equation (12.54). We solve it again, now with both Simpson methods 1 and 2, doing so on $[0, 10]$ with $h = 0.1$. A plot of the error when using Simpson method 1 is given in Figure 12.1, and that for Simpson method 2 is given in Figure 12.2. The error with Simpson method 1 is as predicted from the above discussion: it increases rapidly with increasing t, and it is oscillatory in sign. With Simpson method 2 there is a much more regular and better behavior in the error, in this case of sinusoidal form, reflecting the sinusoidal form of the true solution $Y(t) = \tfrac{1}{2} \left(\cos t - \sin t + e^{-t} \right)$. There are also some oscillations, but they are more minor and are imposed on the dominant form of the error. ∎

A very good introduction to the topic of numerical stability for solving Volterra integral equations is given by Linz [59, Section 7.4]. It also is a very good introduction to the general subject of the numerical solution of Volterra integral equations. An excellent, more recent, and more specialized treatment is given by Brunner [17].

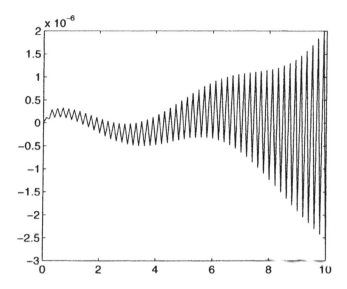

Figure 12.1   The error in solving (12.54) using Simpson method 1

PROBLEMS

1. For the following Volterra integral equations of the second kind, show that the given function $Y(t)$ is the solution of the given equation.

(a)

$$Y(t) = \cos(t) - \int_0^t (t-s) \cos(t-s) Y(s)\, ds,$$

$$Y(t) = \tfrac{2}{3} \cos(\sqrt{3}\, t) + \tfrac{1}{3}.$$

(b)

$$Y(t) = t + \int_0^t \sin(t-s) Y(s)\, ds,$$

$$Y(t) = t + \tfrac{1}{6} t^3.$$

(c)

$$Y(t) = \sinh(t) - \int_0^t \cosh(t-s) Y(s)\, ds,$$

$$Y(t) = \frac{2}{\sqrt{5}} \sinh\left(\frac{\sqrt{5}}{2} t\right) e^{-t/2}.$$

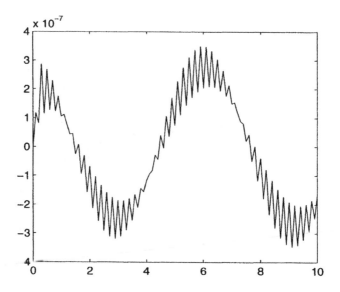

Figure 12.2   The error in solving (12.54) using Simpson method 2

2. Reduce equation (12.15) to (12.12) by introducing the new unknown function $Z(t) = e^{-\lambda t} Y(t)$. Use this transformation to obtain (12.16) from (12.14).

3. Demonstrate formula (12.4).

4. Using mathematical induction, show that the kernels $K_j(t, s)$ of (12.6) satisfy

$$|K_j(t, s)| \leq \frac{(t - s)^{j-1}}{(j - 1)!} B^j, \quad j \geq 1.$$

From this, show that

$$\left| \int_0^t K_j(t, s)\, g(s)\, ds \right| \leq \frac{(tB)^j}{j!} \max_{0 \leq s \leq t} |g(s)|.$$

5. Using the result of Problem 4, and motivated by (12.5), show that the series

$$g(t) + \sum_{j=1}^{\infty} \int_0^t K_j(t, s)\, g(s)\, ds$$

is absolutely convergent. Note that it still remains necessary to show that this function satisfies (12.2). We refer to Linz [59, p. 30] for a proof, along with a proof of the uniquess of the solution.

6. Assume that it has been shown, based on (12.5), that

$$Y(t) = g(t) + \sum_{j=1}^{\infty} \int_0^t K_j(t, s) \, g(s) \, ds$$

is an absolutely convergent series. Combine this with Problem 4 to show that $Y(t)$ satisfies (12.7).

7. Let $Y(t)$ be the continuous solution of (12.2).

   (a) Assume that $K(t, s)$ is differentiable with respect to t and that $\partial K(t, s)/\partial t$ is continuous for $0 \le s \le t \le T$. Assume further that $g(t)$ is continously differentiable on $[0, T]$. Show that $Y(t)$ is differentiable and that

   $$Y'(t) = g'(t) + K(t, t) Y(t) + \int_0^t \frac{\partial K(t, s)}{\partial t} Y(t) \, dt.$$

   (b) Give a corresponding result that guarantees that $Y(t)$ is twice continuously differentiable on $[0, T]$.

8. Using the MATLAB program `vie_trap`, solve (12.23) on $[0, 12]$. Do so for stepsizes $h = 0.2, 0.1, 0.05$; then graph the errors over the full interval.

9. Apply the MATLAB program `vie_trap` to the equation

   $$Y(t) = g(t) + \lambda \int_0^t Y(s) \, ds, \qquad t \ge 0,$$

   $$g(t) = (1 - \lambda) \sin t + (1 + \lambda) \cos t - \lambda$$

   over the interval $[0, 2\pi]$. The true solution is $Y(t) = \cos t + \sin t$. Do so for stepsizes of $h = 0.5, 0.25, 0.125$ and $\lambda = -1, 1$. Observe the decrease in the error as h is halved. Comment on any differences observed between the cases of $\lambda = -1$ and $\lambda = 1$.

10. Using mathematical induction on n, prove Lemma 12.6.

11. In Example 12.8 it is asserted that $Y(t_1) - y_1 = \mathcal{O}(h^3)$. Explain why this is true.

12. Write MATLAB programs for both Simpson methods 1 and 2. Generate y_1 using the trapezoidal method. After writing the program, use it to solve the linear integral equation (12.54), say on $[0, 10]$. Use a stepsize of $h = 0.2$ and graph the errors using MATLAB.

13. Using the programs of Problem 12, solve the equation given in Problem 9. Do so with both Simpson methods. Do so with both $\lambda = -1$ and $\lambda = 1$. Use $h = 0.2, 0.1$ and solve the equation on $[0, 10]$.

14. In analogy with the formulas (12.26)–(12.28) for the quadrature error for the trapezoidal rule, give the corresponding formulas for Simpson method 2. Note that this includes the Newton $\frac{3}{8}$'s rule.

APPENDIX A

TAYLOR'S THEOREM

For a function with a number of derivatives at a specific point, Taylor's theorem provides a polynomial that is close to the function in a neighborhood of the point and an error formula for the difference between the function and the polynomial. Taylor's theorem is an important tool in developing numerical methods and deriving error bounds. We start with a review of the mean value theorem.

Theorem A.1 *(Mean value theorem) Assume that $f(x)$ is continuous on $[a, b]$ and is differentiable on (a, b). Then there is a point $c \in (a, b)$ such that*

$$f(b) - f(a) = f'(c)(b - a). \tag{A.1}$$

The number c in (A.1) is usually unknown. There is an analogous form of the theorem for integrals. Assume that $f(x)$ is continuous on $[a, b]$, $w(x)$ is nonnegative and integrable on $[a, b]$. Then there exists $c \in (a, b)$ for which

$$\int_a^b f(x) \, w(x) \, dx = f(c) \int_a^b w(x) \, dx. \tag{A.2}$$

Theorem A.2 *(Taylor's theorem for functions of one real variable) Assume that $f(x)$ has $n + 1$ continuous derivatives for $a \leq x \leq b$, and let $x_0 \in [a, b]$. Then*

$$f(x) = p_n(x) + R_n(x), \quad a \leq x \leq b, \tag{A.3}$$

where

$$
\begin{aligned}
p_n(x) &= f(x_0) + (x - x_0)f'(x_0) \\
&\quad + \frac{(x - x_0)^2}{2!}f''(x_0) + \cdots + \frac{(x - x_0)^n}{n!}f^{(n)}(x_0) \\
&= \sum_{j=0}^{n} \frac{(x - x_0)^j}{j!} f^{(j)}(x_0)
\end{aligned}
\tag{A.4}
$$

is the Taylor polynomial of degree n for the function $f(x)$ and the point of approximation x_0, and $R_n(x)$ is the remainder in approximating $f(x)$ by $p_n(x)$. We have

$$R_n(x) = \frac{1}{n!} \int_{x_0}^{x} (x - t)^n f^{(n+1)}(t)\, dt \tag{A.5}$$

$$= \frac{(x - x_0)^{n+1}}{(n + 1)!} f^{(n+1)}(c_x) \tag{A.6}$$

with c_x an unknown point between x_0 and x.

The Taylor polynomial is constructed by requiring

$$p_n^{(j)}(x_0) = f^{(j)}(x_0), \quad j = 0, 1, \ldots, n.$$

Thus, we expect $p_n(x)$ is close to $f(x)$, at least for x close to x_0. Two forms of the remainder $R_n(x)$ are given in the theorem. The form (A.6) is derived from (A.5) by an application of the integral form of the mean value theorem, (A.2). The remainder formula (A.5) does not involve an unknown point, and it is useful where precise error bound is needed. In most contexts, the remainder formula (A.6) is sufficient.

Taylor's theorem can be proved by repeated application of the formula

$$g(x) = g(x_0) + \int_{x_0}^{x} g'(t)\, dt \tag{A.7}$$

for a continuously differentiable function g. Evidently, this formula corresponds to Taylor's theorem with $n = 0$. As an example, we illustrate the derivation of (A.3) with $n = 1$; the derivation of (A.3) for $n > 1$ can be done similarly through an inductive argument. We apply (A.7) for $g = f'$:

$$f'(t) = f'(x_0) + \int_{x_0}^{t} f''(s)\, ds.$$

Thus,

$$f(x) = f(x_0) + \int_{x_0}^{x} f'(t) \, dt$$

$$= f(x_0) + \int_{x_0}^{x} \left[f'(x_0) + \int_{x_0}^{t} f''(s) \, ds \right] dt$$

$$= f(x_0) + f'(x_0)(x - x_0) + \int_{x_0}^{x} \int_{x_0}^{t} f''(s) \, ds \, dt.$$

Interchanging the order of integration, we can rewrite the last term as

$$\int_{x_0}^{x} \int_{s}^{x} f''(s) \, dt \, ds = \int_{x_0}^{x} (x - s) f''(s) \, ds.$$

Changing s into t, we have thus shown Taylor's theorem with $n = 1$.

In applying Taylor's theorem, we often need to choose a value for the nonnegative integer n. If we want to have a linear approximation of twice continuously differentiable function $f(x)$ near $x = x_0$, then we take $n = 1$ and write

$$f(x) = f(x_0) + (x - x_0) f'(x_0) + \tfrac{1}{2}(x - x_0)^2 f''(c)$$

for some c between x and x_0. To show that $(f(x + h) - f(x))/h$ ($h > 0$, usually small) is a first-order approximation of $f'(x)$, we choose $n = 1$,

$$f(x + h) = f(x) + h f'(x) + \tfrac{1}{2}h^2 f''(c),$$

and so

$$\frac{f(x + h) - f(x)}{h} = f'(x) + \tfrac{1}{2}h f''(c).$$

As a further example, let us show that $(f(x + h) - f(x))/h$ is a second-order approximation of $f'(x + h/2)$. We choose $n = 2$, and write (here $x_0 = x + \tfrac{1}{2}h$)

$$f(x + h) = f\left(x + \tfrac{1}{2}h\right) + \tfrac{1}{2}h f'\left(x + \tfrac{1}{2}h\right) + \tfrac{1}{2}\left(\tfrac{1}{2}h\right)^2 f''\left(x + \tfrac{1}{2}h\right)$$
$$+ \tfrac{1}{6}\left(\tfrac{1}{2}h\right)^3 f'''(c_1),$$

$$f(x) = f\left(x + \tfrac{1}{2}h\right) - \tfrac{1}{2}h f'(x + h/2) + \tfrac{1}{2}\left(\tfrac{1}{2}h\right)^2 f''(x + h/2)$$
$$- \tfrac{1}{6}\left(\tfrac{1}{2}h\right)^3 f'''(c_2)$$

for some $c_1 \in (x + \tfrac{1}{2}h, x + h)$ and $c_2 \in (x, x + \tfrac{1}{2}h)$. Thus,

$$\frac{f(x + h) - f(x)}{h} = f'\left(x + \tfrac{1}{2}h\right) + \tfrac{1}{48}h^2 \left[f'''(c_1) + f'''(c_2) \right]$$

showing $(f(x + h) - f(x))/h$ is a second-order approximation of $f'(x + \tfrac{1}{2}h)$. This result is usually expressed by saying that $(f(x + h) - f(x - h))/(2h)$ is a second-order approximation to $f'(x)$. Of course, in these preceding examples, we assume the function $f(x)$ has the required number of derivatives.

Sample formulas resulted from Taylor's theorem are

$$e^x = 1 + x + \frac{x^2}{2!} + \cdots + \frac{x^n}{n!} + \frac{x^{n+1}}{(n+1)!}e^c,$$

$$\sin(x) = x - \frac{x^3}{3!} + \frac{x^5}{5!} - \cdots + (-1)^{n-1}\frac{x^{2n-1}}{(2n-1)!} + (-1)^n\frac{x^{2n+1}}{(2n+1)!}\cos(c),$$

$$\cos(x) = 1 - \frac{x^2}{2!} + \frac{x^4}{4!} - \cdots + (-1)^n\frac{x^{2n}}{(2n)!} + (-1)^{n+1}\frac{x^{2n+2}}{(2n+2)!}\cos(c),$$

$$\log(1-x) = -\left(x + \frac{1}{2}x^2 + \cdots + \frac{1}{n+1}x^{n+1}\right) - \left(\frac{1}{1-c}\right)\frac{x^{n+2}}{n+2}, \quad -1 \le x < 1,$$

where c is between $x_0 = 0$ and x. The first three formulas are valid for any $-\infty < x < \infty$.

Theorem A.3 *(Taylor's theorem for functions of two real variables) Assume that $f(x, y)$ has continuous partial derivatives up to order $n + 1$ for $a \le x \le b$ and $c \le y \le d$, and let $x_0 \in [a, b]$, $y_0 \in [c, d]$. Then*

$$f(x, y) = p_n(x, y) + R_n(x, y), \quad a \le x \le b, \ c \le y \le d, \tag{A.8}$$

where

$$p_n(x, y) = f(x_0, y_0)$$
$$+ \sum_{j=1}^{n}\frac{1}{j!}\left[(x - x_0)\frac{\partial}{\partial x} + (y - y_0)\frac{\partial}{\partial y}\right]^j f(x_0, y_0), \tag{A.9}$$

$$R_n(x, y) = \frac{1}{(n+1)!}\left[(x - x_0)\frac{\partial}{\partial x} + (y - y_0)\frac{\partial}{\partial y}\right]^{n+1}$$
$$\times f(x_0 + \theta(x - x_0), y_0 + \theta(y - y_0)) \tag{A.10}$$

with an unknown number $\theta \in (0, 1)$.

In (A.9) and (A.10), the expression

$$\left[(x - x_0)\frac{\partial}{\partial x} + (y - y_0)\frac{\partial}{\partial y}\right]^j f(x_0, y_0)$$

$$= \sum_{i=0}^{j}\frac{j!}{i!(j-i)!}(x - x_0)^i(y - y_0)^{j-i}\frac{\partial^j}{\partial x^i \partial y^{j-i}}f(x_0, y_0)$$

is defined formally through the binomial expansion for numbers:

$$(a + b)^j = \sum_{i=0}^{j}\frac{j!}{i!(j-i)!}a^i b^{j-i}.$$

For example, with $j = 2$, we obtain

$$\left[(x - x_0)\frac{\partial}{\partial x} + (y - y_0)\frac{\partial}{\partial y} \right]^2 f(x_0, y_0)$$

$$= (x - x_0)^2 \frac{\partial^2}{\partial x^2} f(x_0, y_0) + 2(x - x_0)(y - y_0)\frac{\partial^2}{\partial x \partial y} f(x_0, y_0)$$

$$+ (y - y_0)^2 \frac{\partial^2}{\partial y^2} f(x_0, y_0).$$

Formula (A.8) with (A.9)–(A.10) can be proved by applying Taylor's theorem for one real variable as follows. Define a function of one real variable

$$F(t) = f(x_0 + t(x - x_0), y_0 + t(y - y_0)).$$

Note that $F(0) = f(x_0, y_0)$, $F(1) = f(x, y)$. Applying formula (A.3) with (A.4) and (A.6), we obtain

$$F(1) = F(0) + \sum_{j=1}^{n} \frac{1}{j!} F^{(j)}(0) + \frac{1}{(n+1)!} F^{(n+1)}(\theta)$$

for some unknown number $\theta \in (0, 1)$. Using the chain rule, we can verify that

$$F^{(j)}(0) = \left[(x - x_0)\frac{\partial}{\partial x} + (y - y_0)\frac{\partial}{\partial y} \right]^j f(x_0, y_0).$$

This argument is also valid when the function has m $(m > 2)$ real variables, leading to Taylor's theorem for functions of m real variables.

APPENDIX B

POLYNOMIAL INTERPOLATION

The problem of polynomial interpolation is the selection of a particular polynomial $p(x)$ from a given class of polynomials in such a way that the graph of $y = p(x)$ passes through a finite set of given data points. Polynomial interpolation theory has many important uses, but in this text we are interested in it primarily as a tool for developing numerical methods for solving ordinary differential equations.

Let $x_0, x_1, \ldots, x_n$ be distinct real or complex numbers, and let $y_0, y_1, \ldots, y_n$ be associated function values. We now study the problem of finding a polynomial $p(x)$ that interpolates the given data:

$$p(x_i) = y_i, \quad i = 0, 1, \ldots, n. \tag{B.1}$$

Does such a polynomial exist, and if so, what is its degree? Is it unique? What formula can we use to for produce $p(x)$ from the given data?

By writing

$$p(x) = a_0 + a_1 x + \cdots + a_m x^m$$

for a general polynomial of degree m, we see that there are $m + 1$ independent parameters $a_0, a_1, \ldots, a_m$. Since (B.1) imposes $n + 1$ conditions on $p(x)$, it is reasonable to first consider the case when $m = n$. Then we want to find $a_0, a_1, \ldots, a_n$

such that

$$a_0 + a_1 x_0 + a_2 x_0^2 + \cdots + a_n x_0^n = y_0,$$

$$\vdots$$

$$a_0 + a_1 x_n + a_2 x_n^2 + \cdots + a_n x_n^n = y_n. \tag{B.2}$$

This is a system of $n + 1$ linear equations in $n + 1$ unknowns, and solving it is completely equivalent to solving the polynomial interpolation problem. In vector–matrix notation, the system is

$$X a = y$$

with

$$X = \begin{bmatrix} 1 & x_0 & x_0^2 & \cdots & x_0^n \\ \vdots & & & & \vdots \\ 1 & x_{n-1} & x_{n-1}^2 & \cdots & x_{n-1}^n \\ 1 & x_n & x_n^2 & \cdots & x_n^n \end{bmatrix}, \tag{B.3}$$

$$a = [a_0, a_1, \ldots, a_n]^T, \qquad y = [y_0, \ldots, y_n]^T.$$

The matrix X is called a *Vandermonde matrix*, and its determinant is given by

$$\det(X) = \prod_{0 \le j < i \le n} (x_i - x_j).$$

Theorem B.1 *Given $n + 1$ distinct points $x_0, \ldots, x_n$ and $n + 1$ ordinates $y_0, \ldots, y_n$, there is a polynomial $p(x)$ of degree $\le n$ that interpolates y_i at x_i, $i = 0, 1, \ldots, n$. This polynomial $p(x)$ is unique in the set of all polynomials of degree $\le n$.*

Proof. There are a number of different proofs of this important result. We give a constructive proof that exhibits explicitly the interpolating polynomial $p(x)$ in a form useful for the applications in this text.

To begin, consider the special interpolation problem in which

$$y_i = 1, \qquad y_j = 0 \qquad \text{for } j \ne i$$

for some i, $0 \le i \le n$. We want a polynomial of degree $\le n$ with the n zeros x_j, $j \ne i$. Then

$$p(x) = c(x - x_0) \cdots (x - x_{i-1})(x - x_{i+1}) \cdots (x - x_n)$$

for some constant c. The condition $p(x_i) = 1$ implies

$$c = [(x_i - x_0) \cdots (x_i - x_{i-1})(x_i - x_{i+1}) \cdots (x_i - x_n)]^{-1}.$$

This special polynomial is written as

$$l_i(x) = \prod_{j \ne i} \left(\frac{x - x_j}{x_i - x_j} \right), \qquad i = 0, 1, \ldots, n. \tag{B.4}$$

To solve the general interpolation problem (B.1), we can write

$$p(x) = y_0 l_0(x) + y_1 l_1(x) + \cdots + y_n l_n(x).$$

With the special properties of the polynomials $l_i(x)$, it is easy to show that $p(x)$ satisfies (B.1). Also, degree $p(x) \leq n$ since all $l_i(x)$ have degree n.

To prove uniqueness, suppose that $q(x)$ is another polynomial of degree $\leq n$ that satisfies (B.1). Define

$$r(x) = p(x) - q(x).$$

Then degree $r(x) \leq n$ and

$$r(x_i) = p(x_i) - q(x_i) = y_i - y_i = 0, \qquad i = 0, 1, \ldots, n.$$

Since $r(x)$ has $n + 1$ zeros, we must have $r(x) \equiv 0$. This proves $p(x) \equiv q(x)$. ∎

The formula

$$p_n(x) = \sum_{i=0}^{n} y_i l_i(x) \tag{B.5}$$

is called *Lagrange's formula* for the interpolating polynomial.

Example B.2

$$p_1(x) = \frac{x - x_1}{x_0 - x_1} y_0 + \frac{x - x_0}{x_1 - x_0} y_1 = \frac{(x_1 - x)y_0 + (x - x_0)y_1}{x_1 - x_0},$$

$$p_2(x) = \frac{(x - x_1)(x - x_2)}{(x_0 - x_1)(x_0 - x_2)} y_0 + \frac{(x - x_0)(x - x_2)}{(x_1 - x_0)(x_1 - x_2)} y_1 + \frac{(x - x_0)(x - x_1)}{(x_2 - x_0)(x_2 - x_1)} y_2.$$

The polynomial of degree ≤ 2 that passes through the three points $(0, 1)$, $(-1, 2)$, and $(1, 3)$ is

$$p_2(x) = \frac{(x + 1)(x - 1)}{(0 + 1)(0 - 1)} \cdot 1 + \frac{(x - 0)(x - 1)}{(-1 - 0)(-1 - 1)} \cdot 2 + \frac{(x - 0)(x + 1)}{(1 - 0)(1 + 1)} \cdot 3$$

$$= 1 + \tfrac{1}{2}x + \tfrac{3}{2}x^2. \qquad \blacksquare$$

If a function $f(x)$ is given, then we can form an approximation to it using the interpolating polynomial

$$p_n(x; f) \equiv p_n(x) = \sum_{i=0}^{n} f(x_i) l_i(x). \tag{B.6}$$

This interpolates $f(x)$ at $x_0, \ldots, x_n$. This polynomial formula is used at several points in this text.

The basic result used in analyzing the error of interpolation is the following theorem. As a notation, $\mathcal{H}\{a, b, c, \ldots\}$ denotes the smallest interval containing all of the real numbers $a, b, c, \ldots$.

Theorem B.3 *Let $x_0, x_1, \ldots, x_n$ be distinct real numbers, and let f be a real valued function with $n + 1$ continuous derivatives on the interval $I_t = \mathcal{H}\{t, x_0, \ldots, x_n\}$ with t some given real number.*
Then there exists $\xi \in I_t$ with

$$f(t) - \sum_{j=0}^{n} f(x_j) l_j(t) = \frac{(t - x_0) \cdots (t - x_n)}{(n + 1)!} f^{(n+1)}(\xi). \tag{B.7}$$

A proof of this result can be found in many numerical analysis textbooks; e.g., see [11, p. 135]. The theory and practice of polynomial interpolation represent a very large subject. Again, most numerical analysis textbooks contain a basic introduction, and we refer the interested reader to them.

REFERENCES

1. R. Aiken (editor). *Stiff Computation*, Oxford University Press, Oxford, 1985.

2. R. Alexander. "Diagonally implicit Runge-Kutta methods for stiff ODE's", *SIAM Journal on Numerical Analysis* **14** (1977), pp. 1006–1021.

3. E. Anderson, Z. Bai, C. Bischof, J. Demmel, J. Dongarra, J. DuCroz, A. Greenbaum, S. Hammarling, A. McKenney, S. Ostrouchov, and D. Sorenson. *LAPACK Users' Guide*, SIAM Pub., Philadelphia, 1992.

4. V. Arnold. *Mathematical Methods of Classical Mechanics*, Springer–Verlag, New York, 1974.

5. U. Ascher, H. Chin, and S. Reich. "Stabilization of DAEs and invariant manifolds", *Numerische Mathematik* **67** (1994), pp. 131–149.

6. U. Ascher, J. Christiansen, and R. Russell. "Collocation software for boundary-value ODEs", *ACM Trans. Math. Soft.* **7** (1981), pp. 209–222.

7. U. Ascher, J. Christiansen, and R. Russell. "COLSYS: Collocation software for boundary-value ODEs", *ACM Trans. Math. Soft.* **7** (1981), pp. 223–229.

8. U. Ascher and R. Russell, eds. *Numerical Boundary Value ODEs*, Birkhäuser, Boston, MA, 1985.

9. U. Ascher, R. Mattheij, and R. Russell. *Numerical Solution of Boundary Value Problems for Ordinary Differential Equations*, Prentice-Hall, Englewood Cliffs, New Jersey, 1988.

10. U. Ascher and L. Petzold. *Computer Methods for Ordinary Differential Equations and Differential-Algebraic Equations*, SIAM, Philadelphia, 1998.

245

11. K. Atkinson. *An Introduction to Numerical Analysis*, 2nd ed., John Wiley, New York, 1989.

12. K. Atkinson and W. Han. *Elementary Numerical Analysis*, 3rd ed., John Wiley, New York, 2004.

13. A. Aziz. *Numerical Solutions of Boundary Value Problems for Ordinary Differential Equations*, Academic Press, New York, 1975.

14. C. Baker. *The Numerical Treatment of Integral Equations*, Clarendon Press, Oxford, 1977.

15. J. Baumgarte. "Stabilization of constraints and integrals of motion in dynamical systems", *Computer Methods in Applied Mechanics and Engineering* **1** (1972), pp. 1–16.

16. W. Boyce and R. DiPrima. *Elementary Differential Equations*, 7th edition, John Wiley & Sons, 2003.

17. H. Brunner. *Collocation Methods for Volterra Integral and Related Functional Equations*, Cambridge Univ. Press, 2004.

18. P. Bogacki and L. Shampine. "A 3(2) pair of Runge-Kutta formulas", *Appl. Math. Lett.* **2** (1989), pp. 321–325.

19. K.E. Brenan, S.L. Campbell and L.R. Petzold. *Numerical Solution of Initial-Value Problems in Differential-Algebraic Equations*, Number 14 in Classics in Applied Mathematics. SIAM Publ., Philadelphia, PA, 1996. Originally published by North Holland, 1989.

20. K.E. Brenan and B.E. Engquist. "Backward differentiation approximations of nonlinear differential/algebraic systems", *Mathematics of Computation* **51** (1988), pp. 659–676.

21. P.N. Brown, A.C. Hindmarsh, and L.R. Petzold. "Using Krylov methods in the solution of large-scale differential-algebraic systems", *SIAM J. Scientific Computing* **15** (1994), pp. 1467–1488.

22. K. Burrage and J.C. Butcher. "Stability criteria for implicit Runge-Kutta methods", *SIAM J. Numer. Anal.* **16** (1979), pp. 46–57.

23. J.C. Butcher. "Implicit Runge-Kutta processes", *Math. Comp.* **18** (1964), pp. 50–64.

24. J.C. Butcher. "A stability property of implicit Runge–Kutta methods", *BIT* **15** (1975), pp. 358–361.

25. J.C. Butcher. "General linear methods", *Acta Numerica* **15** (2006), Cambridge University Press.

26. S.L. Campbell, R. Hollenbeck, K. Yeomans and Y. Zhong. "Mixed symbolic-numerical computations with general DAEs. I. System properties", *Numerical Algorithms* **19** (1998), pp. 73–83.

27. J. Cash. "On the numerical integration of nonlinear two-point boundary value problems using iterated deferred corrections. II. The development and analysis of highly stable deferred correction formulae", *SIAM J. Numer. Anal.* **25** (1988), pp. 862–882.

28. B. Childs, E. Denman, M. Scott, P. Nelson, and J. Daniel, eds. *Codes for Boundary-Value Problems in Ordinary Differential Equations*, Lec. Notes in Comp. Sci. **76**, Springer-Verlag, New York, 1979.

29. G.F. Corliss, A. Griewank, P. Henneberger, G. Kirlinger, F.A. Potra, and H.J. Stetter. "High-order stiff ODE solvers via automatic differentiation and rational prediction", in *Numerical Analysis and its Applications* (Rousse, 1996), Lecture Notes in Computer Science **1196**, pp. 114–125. Springer–Verlag, Berlin, 1997.

30. M. Crouzeix. "Sur la *B*-stabilité des méthodes de Runge-Kutta". *Numer. Math.* **32** (1979), pp. 75–82.

31. M. Crouzeix and P.A. Raviart. "Approximation des problèmes d'évolution", unpublished lecture notes, Université de Rennes, 1980.

32. P. Deuflhard. "Nonlinear equation solvers in boundary value problem codes", in *Codes for Boundary-Value Problems in Ordinary Differential Equations*, B. Childs, M. Scott, J. Daniel, E. Denman, and P. Nelson, eds., Lec. Notes in Comp. Sci. **76**, Springer-Verlag, New York, 1979, pp. 40–66.

33. P. Deuflhard and F. Bornemann. *Scientific Computing with Ordinary Differential Equations*, Springer-Verlag, 2002.

34. J. Dormand and P. Prince."A family of embedded Runge-Kutta formulae", *J. Comp. Appl. Math.* **6** (1980), pp. 19–26.

35. W. Enright. "Improving the performance of numerical methods for two-point boundary value problems", in *Numerical Boundary Value ODEs*, U. Ascher and R. Russell, eds., Birkhäuser, Boston, MA, 1985, pp. 107–119.

36. K. Eriksson, D. Estep, P. Hansbo, and C. Johnson. "Introduction to adaptive methods for differential equations", *Acta Numerica* **5** (1995), Cambridge University Press.

37. A. Fasano and S. Marmi. *Analytical Mechanics: An Introduction*. Oxford University Press, Oxford, 2006.

38. G.R. Fowles. *Analytical Mechanics*, Holt, Rinehart and Winston, 1962.

39. C. W. Gear. *Numerical Initial Value Problems in Ordinary Differential Equations*, Prentice-Hall, Englewood Cliffs, NJ, 1971.

40. C.W. Gear, B. Leimkuhler, and G.K. Gupta. "Automatic integration of Euler–Lagrange equations with constraints", in *Proceedings of the International Conference on Computational and Applied Mathematics* (Leuven, 1984), Vol. **12/13** (1985), pp. 77–90.

41. I. Gladwell and D. Sayers. *Computational Techniques for Ordinary Differential Equations*, Academic Press, New York, 1980.

42. E. Hairer, C. Lubich, and M. Roche. *The Numerical Solution of Differential-Algebraic Systems by Runge–Kutta Methods*. Lecture Notes in Mathematics **1409** (1989), Springer–Verlag, Berlin.

43. E. Hairer, C. Lubich, and G. Wanner. "Geometric numerical integration illustrated by the Störmer-Verlet method", *Acta Numerica* **12** (2003), Cambridge University Press.

44. E. Hairer and G. Wanner. *Solving Ordinary Differential Equations. II. Stiff and Differential-Algebraic Problems*, 2nd ed., Springer-Verlag, Berlin, 1996.

45. P. Henrici. *Discrete Variable Methods in Ordinary Differential Equations*, John Wiley, 1962.

46. A. Hindmarsh, P. Brown, K. Grant, S. Lee, R. Serban, D. Shumaker, and C. Woodward. SUNDIALS: Suite of Nonlinear and Differential/Algebraic Equation Solvers, *ACM Transactions on Mathematical Software* **31** (2005), pp. 363–396. Also, go to the URL https://computation.llnl.gov/casc/sundials/

47. E. Isaacson and H. Keller. *Analysis of Numerical Methods*, John Wiley, New York, 1966.

48. A. Iserles. *A First Course in the Numerical Analysis of Differential Equations*, Cambridge University Press, Cambridge, United Kingdom, 1996.

49. L. Jay. "Convergence of Runge-Kutta methods for differential-algebraic systems of index 3", *Applied Numerical Mathematics* **17** (1995), pp. 97–118.

50. L. Jay. "Symplectic partitioned Runge-Kutta methods for constrained Hamiltonian systems", *SIAM Journal on Numerical Analysis* **33** (1996), pp. 368–387.

51. L. Jay. "Specialized Runge-Kutta methods for index 2 differential-algebraic equations", *Mathematics of Computation* **75** (2006), pp. 641–654.

52. H. Keller. *Numerical Solution of Two-Point Boundary Value Problems*, Regional Conf. Series in Appl. Maths. **24**, SIAM Pub., Philadelphia, PA, 1976.

53. H. Keller. *Numerical Methods for Two-Point Boundary Value Problems*, Dover, New York, 1992 (corrected reprint of the 1968 edition, Blaisdell, Waltham, MA).

54. H. Keller and S. Antman, eds. *Bifurcation Theory and Nonlinear Eigenvalue Problems*, Benjamin, New York, 1969.

55. C.T. Kelley. *Solving Nonlinear Equations with Newton's Method*, SIAM Pub., Philadelphia, 2003.

56. W. Kelley and A. Peterson. *Difference Equations*, 2nd ed., Academic Press, Burlington, Massachusetts, 2001.

57. R. Kress. *Numerical Analysis*, Springer-Verlag, New York, 1998.

58. J. Lambert. *Computational Methods in Ordinary Differential Equations*, John Wiley, New York, 1973.

59. P. Linz. *Analytical and Numerical Methods for Volterra Equations*, SIAM Pub., 1985.

60. P. Lötstedt and L. Petzold. "Numerical solution of nonlinear differential equations with algebraic constraints. I. Convergence results for backward differentiation formulas", *Mathematics of Computation* **46** (1986), pp. 491–516.

61. J. Marsden and T. Ratiu. *Introduction to Mechanics and Symmetry*, Springer-Verlag, New York, 1999.

62. R. März. "Numerical methods for differential algebraic equations", *Acta Numerica 1992*, Cambridge University Press, 1992.

63. D. Melgaard and R. Sincovec. "Algorithm 565: PDETWO/PSETM/GEARB: Solution of systems of two-dimensional nonlinear partial differential equations", *ACM Trans. Math. Software* **7** (1981), pp. 126–135.

64. R. Miller. *Nonlinear Volterra Integral Equations*, Benjamin Pub., 1971.

65. L.R. Petzold. "A description of DASSL: A differential-algebraic system solver", in R. S. Stepleman, editor, *Scientific Computing*, pp. 65–68. North-Holland, Amsterdam, 1983.

66. L. Petzold, L. Jay, and J. Yen. "Numerical solution of highly oscillatory ordinary differential equations", *Acta Numerica* **6** (1997), Cambridge University Press.

67. E. Platen. "An introduction to numerical methods for stochastic differential equations", *Acta Numerica* **8** (1999), Cambridge University Press.

68. A. Quarteroni, R. Sacco, and F. Saleri. *Numerical Mathematics*, Springer-Verlag, New York, 2000.

69. L.B. Rall and G.F. Corliss. "An introduction to automatic differentiation", in *Computational Differentiation* (Santa Fe, NM, 1996), pp. 1–18. SIAM, Philadelphia, PA, 1996.

70. J. Sanz-Serna. "Symplectic integrators for Hamiltonian problems: an overview", *Acta Numerica 1992*, Cambridge University Press, 1992.

71. W. Schiesser. *The Numerical Method of Lines*, Academic Press, San Diego, 1991.

72. L. Shampine. *Numerical Solution of Ordinary Differential Equations*, Chapman & Hall, New York, 1994.

73. L. Shampine and M. Reichelt. "The MATLAB ODE Suite", *SIAM Journal on Scientific Computing* **18** (1997), pp. 1–22.

74. L. Shampine, I. Gladwell, and S. Thompson. *Solving ODEs with MATLAB*, Cambridge University Press, 2003.

75. R. Sincovec and N. Madsen. "Software for nonlinear partial differential equations", *ACM Trans. Math. Software* **1** (1975), pp. 232–260.

76. A. Stuart. "Numerical analysis of dynamical systems", *Acta Numerica 1994*, Cambridge University Press, 1994.

77. T. Van Hecke and M. Van Daele. "High-order convergent deferred correction schemes based on parameterized Runge-Kutta-Nyström methods for second-order boundary value problems. Advanced numerical methods for mathematical modelling", *J. Comput. Appl. Math.* **132** (2001), pp. 107–125.

78. D. Widder. *The Heat Equation*, Academic Press, New York, 1975.

INDEX

PURE AND APPLIED MATHEMATICS

A Wiley-Interscience Series of Texts, Monographs, and Tracts

Founded by RICHARD COURANT
Editors Emeriti: MYRON B. ALLEN III, DAVID A. COX, PETER HILTON,
HARRY HOCHSTADT, PETER LAX, JOHN TOLAND

*Now available in a lower priced paperback edition in the Wiley Classics Library.
†Now available in paperback.

*Now available in a lower priced paperback edition in the Wiley Classics Library.
†Now available in paperback.

SEWELL—Computational Methods of Linear Algebra, Second Edition
SHICK—Topology: Point-Set and Geometric
*SIEGEL—Topics in Complex Function Theory
        Volume 1—Elliptic Functions and Uniformization Theory
        Volume 2—Automorphic Functions and Abelian Integrals
        Volume 3—Abelian Functions and Modular Functions of Several Variables
SMITH and ROMANOWSKA—Post-Modern Algebra
ŠOLÍN–Partial Differential Equations and the Finite Element Method
STADE—Fourier Analysis
STAKGOLD—Green's Functions and Boundary Value Problems, Second Editon
STAHL—Introduction to Topology and Geometry
STANOYEVITCH—Introduction to Numerical Ordinary and Partial Differential
        Equations Using MATLAB®
*STOKER—Differential Geometry
*STOKER—Nonlinear Vibrations in Mechanical and Electrical Systems
*STOKER—Water Waves: The Mathematical Theory with Applications
WATKINS—Fundamentals of Matrix Computations, Second Edition
WESSELING—An Introduction to Multigrid Methods
†WHITHAM—Linear and Nonlinear Waves
ZAUDERER—Partial Differential Equations of Applied Mathematics, Third Edition

*Now available in a lower priced paperback edition in the Wiley Classics Library.
†Now available in paperback.

Printed and bound by CPI Group (UK) Ltd, Croydon, CR0 4YY

16/04/2025

14658365-0003